Celestial Signs and Classical Rhetoric
in Early Imperial China

SUNY series in Chinese Philosophy and Culture

Roger T. Ames, editor

Celestial Signs and Classical Rhetoric in Early Imperial China

JESSE J. CHAPMAN

Cover: Excavated manuscript from Mawangdui, a Han tomb in Changsha, China. Courtesy Hunan Provincial Museum, Duan Xiaoming, Director.

Published by State University of New York Press, Albany

© 2025 State University of New York

All rights reserved

Printed in the United States of America

No part of this book may be used or reproduced in any manner without written permission. No part of this book may be stored in a retrieval system or transmitted in any form or by any means including electronic, electrostatic, magnetic tape, mechanical, photocopying, recording, or otherwise without the prior permission in writing of the publisher.

Links to third-party websites are provided as a convenience and for informational purposes only. They do not constitute an endorsement or an approval of any of the products, services, or opinions of the organization, companies, or individuals. SUNY Press bears no responsibility for the accuracy, legality, or content of a URL, the external website, or for that of subsequent websites.

EU GPSR Authorised Representative:
Logos Europe, 9 rue Nicolas Poussin, 17000, La Rochelle, France
contact@logoseurope.eu

For information, contact State University of New York Press, Albany, NY
www.sunypress.edu

Library of Congress Cataloging-in-Publication Data

Name: Chapman, Jesse J., 1979– author.
Title: Celestial signs and classical rhetoric in early imperial China / Jesse J. Chapman.
Description: Albany : State University of New York Press, [2025] | Series: SUNY series in Chinese philosophy and culture | Includes bibliographical references and index.
Identifiers: LCCN 2024027003 | ISBN 9798855800548 (hardcover : alk. paper) | ISBN 9798855800555 (ebook)
Subjects: LCSH: Divination—China—History. | Astronomy—China—History. | Omens—China. | Cognition and culture—China—History. | China—Civilization—221 B.C.–960 A.D.
Classification: LCC BL613 .C44 2025 | DDC 299.5/1132—dc23/eng/20241114
LC record available at https://lccn.loc.gov/2024027003

This volume is dedicated to my parents,

*James Martin Chapman (1946–2024) and
Patricia Mae Chapman (1952–),*

who lined our walls with books.

Contents

Acknowledgments ix

Introduction: Celestial Signs, *Tianwen*, and the Emergence of Classicism in Early China 1

Part I
Technical Manuals and Classical Authority

Chapter 1 Practical Omenology and Classical Ideals at Mawangdui 21

Chapter 2 Sign-Reading and Sagely Rulership in the *Huainanzi* and *Shiji* 49

Chapter 3 Integrating Classical Texts with Omen-Reading Technology 79

Part II
Rhetoric in Practice

Chapter 4 Classical Authority and the Elimination of Baleful Signs 105

Chapter 5 Reading Signs, Reading the World 141

Conclusion: Interwoven Discourses and Contextualized Interpretation   173

Appendix A: Terminology and Translation   179

Appendix B: The Bibliographic Category of *Tianwen*   185

Notes   187

Works Cited   235

Index   249

Acknowledgments

This project has been made possible through the engagement and encouragement of an outstanding intellectual community, generous financial support, and the love (and tolerance) of my family and personal friends.

I want to express my profound gratitude to my mentors, Mark Csikszentmihalyi and Michael Nylan, who challenged me to ask deep, interdisciplinary questions and trained me in the skills needed to pursue them. Along with Zhou Yiqun at Stanford, they have also encouraged me to keep going in the face of difficult circumstances. This book would not exist without their support. I owe also a great debt to Robert Ashmore, who taught me methods for reading Chinese poetry that influenced my understanding of several of the figures in this book, though a study of their poetic works must be left for another project, and to the late Nathan Sivin, who provided me with extensive feedback on my doctoral thesis. Charles Sanft graciously read and commented on materials included in part II of this volume. John Wentworth served as copyeditor, with Catherine Shaw and Amy Chapman providing proofreading support. Comments and advice from anonymous peer reviewers proved invaluable. Conversations and coursework during my graduate school years with Francesca Rochberg, Paula Varsano, Andrew Jones, Terry Kleeman, Patricia Berger, William Schaefer, Alex Cook, Robert Sharf, Nicholas Tackett, Robert Alter, and Massimo Mazzotti likewise shaped aspects of this project. I would also like to thank fellow members of two graduate student dissertation groups supported by the Walter and Elise Haas Fund at Berkeley, especially Scott McGinnis. My ongoing collaboration since 2014 with Ralph Neuhäuser, astrophysicist at the Friedrich-Schiller-Universität Jena, and his team has helped me appreciate the implications of understanding celestial signs for modern astronomical purposes. I want to thank James Peltz and Diane Ganeles at SUNY Press, for guiding me through the revision process, and Roger Ames, for managing an interdisciplinary series in which this volume could find a home. He Jianye, Zheng Yifan, and the Multnomah County Interlibrary Loan Office provided crucial research support at times when I was working outside academia.

Early financial support for this work came from the UC Berkeley Graduate Division, Department of East Asian Languages & Cultures, and Center for Chinese Studies. After completing my dissertation, I benefited from a postdoctoral position at the Center for East Asian Studies at Stanford, as well as visiting teaching positions in History at the University of Oklahoma-Norman, UC Merced, and UCLA, in East Asian Studies at NYU, and in EALC at Berkeley. Garrett Olberding, Elyssa Faison, Miriam Gross, Sean Malloy, Laurence Coderre, Richard von Glahn, and Andrea Goldman provided warm welcomes, good conversation, and teaching support along the way. Anti-poverty programs, including the UC Berkeley Parent Grant and the Supplemental Nutrition Assistance Program, have at times made a critical difference.

Friendships, too, have been critical. Nick Deshais has been a constant emotional support. Tobias Zürn has kept me connected to a community of exegetical scholars. Sky, Ian, Dom, Zoe, and their families have, along with everything else, provided me with comfy accommodations while I commuted from Portland to teach in the Bay Area. Brendon, Sunny, Chen, Makoto, and Folkert—through their warmth, intelligence, and mutual curiosity—inspired my interest in language and culture.

Teachers prior to graduate school provided me with foundational knowledge and encouraged a spirit of inquiry that I have endeavored to honor through this work. Jonathan Pease trained me in Classical Chinese and introduced me to the *Yanzi*, Zhang Lihua taught me Mandarin, Karen Phalet pushed me to think beyond the West, and John Hoover and Serna Teixeira encouraged curiosity and rigor in studies of science and of history.

Teaching and conducting research is a wonderful thing to do with your life, if you can manage it, but academia can be hard on families. I am grateful for the lifelong support of my parents, Jim and Patti, my brothers, Hal, Rob, and Michael, my cousin Brandee, my broader extended family, and my in-laws, the Webbs and the Bolivars. My deepest thanks go to my wife, Amy Kate, and my daughters, Ellie, Lola, and Katy, my closest companions on a long and uncertain journey.

Parts of chapters 2 and 3 appear, in a somewhat different and abbreviated format, in my essay "Celestial Signs in Three Historical Treatises," in *Technical Arts in the Han Histories*, eds. Mark Csikszentmihalyi and Michael Nylan (Albany: SUNY Press, 2021), 181–211.

I am grateful to Huang Yijun and Guo Weimin for facilitating communications regarding cover art permissions, and to the Hunan Museum and Director Duan Xiaoming for granting permission to use the cover art image. The image originally appears in Qiu Xigui 裘锡圭 ed. *Changsha Mawangdui Hanmu jianbo jicheng* 長沙馬王堆漢墓簡帛集成, vol. 1 (Beijing: Zhonghua shuju, 2014), 204.

Introduction

Celestial Signs, *Tianwen*, and the Emergence of Classicism in Early China

Eclipses and earthquakes, comets and oddly shaped clouds, halos, rainbows, violent winds, and planets in retrograde motion are natural phenomena that in early China held deep cultural valences. By virtue of the perceived or actual irregularity of their occurrence, the oddity of their appearance, or their power to wreak havoc on agricultural production, such phenomena became objects of ritual supplication, observation and interpretation, and meticulous record-keeping. In the late Western Han (206 BCE–9 CE) and Eastern Han (25 CE–220), erudite classicist writers would engage with such signs across a range of discourses, compiling technical treatises, revisiting historical and legendary narratives, and interpreting omens to call for ritual and administrative reforms.

The ability to interpret and invoke signs comes with power. Those with sufficient erudition could claim to understand the words of a sage, the cause of a flood, or the significance of a strange light in the heavens, and they could employ their knowledge and insight to make authoritative arguments about religion, politics, and society. In Han times, the ability to read and deploy such signs became a core power of classicist court officials. Centered on the rhetorical and religious significance of astronomical and meteorological phenomena in the Western and Eastern Han dynasties, this book investigates how that power came into being and worked in practice.

Jing, the Five Classics, and Classical Texts

Officials versed in classical learning claimed expertise in the interpretation of authoritative texts, wielding that authority for themselves. The authority of the texts and their interpreters emerged through historical processes including both individual acts and institutional changes. Copyists, compilers, and perhaps some composers designated certain texts as "classics" or *jing* 經, a word that also refers to "constancy" or the "warp" of a textile, or verbally means "to manage" or "to pass through."[1] In the early-to-mid Western Han, the label *jing* applied to remarkably varied types of texts, from works of political philosophy, such as the *Daodejing* 道德經 (Classic of the Way and Its Power), to celestial journey poetry, such as the "Li sao jing" 離騷經 (Classic of Encountering Sorrow), though the texts themselves in many cases remained unfixed in orthography, form, and even content.[2] In the late Western Han, reformist thinkers seeking to reinvent Han institutions on the model of those of an idealized early Western Zhou (1046–771) elevated a set of Five Classics—the *Changes* (Yijing 易經), the *Odes* (Shijing 詩經), the *Documents* (Shangshu 尚書), the *Rites* (Li 禮), and the *Annals* (Chunqiu 春秋)—above all others. These texts alone are hereafter referred to as Classics (as opposed to the broader designation *classical texts*). Sign-reading was a central function of two of these Classics, with the *Changes* providing cosmological models that explained how and why milfoil casting worked, and the *Annals* recording cosmic and human events in terse language that later exegetes would mine for sagely judgments.[3] During the same period, the court of Emperor Cheng 成 (r. 33–7 BCE) carried out a massive bibliographical project, gathering texts from throughout the empire, collating, correcting, and cataloging what they found.[4] *Jing* took on a tighter association with the Five Classics, though its older and broader sense remained active. Classicists often specialized in one of these Classics, yet the greatest among them were masters of a wide-ranging textual tradition. This broader body of classical texts included not only the Classics themselves but also traditions or commentaries attached to them, as well as masterworks attributed to Chunqiu (722–481) and Warring States–era (481–222) thinkers. More so than ever before, thinkers and writers integrated ideals, structures, and citations drawn from the Five Classics into explanations for historical change, interpretations of omens, and memorials to the throne.

Perspectives on the Meaningfulness of Omens

This book centers on the reading of a broad subcategory of celestial signs—omens. These include signs appearing in the heavens, signs attributed to the will of Heaven, and other phenomena discussed in technical literature on celestial signs such as wind storms, floods, and earthquakes. Officials claimed knowledge of the etiology of such signs to make arguments concerning what to do about them. One view in early China held that baleful signs, harmful as they were, came about for no particular reason. Bad fortune was just bad luck. A dialogue between the bumbling Duke Jing of Qi 齊景公 (r. 547–490) and his wise chief minister, Yan Ying 晏嬰 (6th cent. BCE), collected in the *Yanzi chunqiu* 晏子春秋 (Annals of Master Yan), demonstrates the contrast. When a baleful sign appeared, the duke sought to use ritual means to remove it. Master Yan instead insisted that the consequences of the sign could be avoided only if the duke reformed his court:

> 日暮，公西面望，睹彗星，召伯常騫，使禳去之。晏子曰：「不可！此天教也。日月之氣，風雨不時，彗星之出，天為民之亂見之，故詔之妖祥，以戒不敬。今君若設文而受諫，謁聖賢人，雖不去彗星，將自亡。」

> At sunset, the duke gazed westward and saw a broom-star (i.e., a comet). He summoned Bo Changqian to exorcise it. Master Yan said, "You must not do so, for this is the Teaching of Heaven. The *qi* around the sun and the moon, unseasonable wind and rain, and the appearance of comets are all produced by Heaven due to disorder among the populace. Inauspicious portents are sent down to warn those who are not sufficiently reverent! Now, if you were to delight in civility (*wen* 文),[5] accept admonitions, and call on sages and worthy men, then even if you did not banish the broom-star, it would disappear on its own."[6]

To Duke Jing, the comet was a harbinger of disaster that appeared for unknown reasons, while for Master Yan, it was a direct result of ritual and political lapses on the part of the ruler. Master Yan aimed to replace the duke's understanding of the comet's etiology with his own.

Different actors interpreted the same signs in different ways. The proper response to baleful signs was open to debate. Duke Jing proposed that a ritual be performed to drive the comet away. Master Yan claimed that the ritual would ultimately fail if the duke did not address the circumstances that had brought the comet into being. Claims regarding why baleful signs appeared underscored arguments regarding the proper response to such signs. Duke Jing showed no concern for the causes behind the comet, but Master Yan spoke as if he knew its precise source.

Others expressed doubt that celestial phenomena were meaningful at all. The late Warring States–era thinker Xunzi 荀子 (ca. 340–ca. 245) insisted that Heaven operated quite apart from human actions or human concerns. Whether a sage-king or a cruel tyrant was in power, the sun, moon, and stars moved in the same way. Floods and droughts came and went without regard to human affairs, unmoved by ritual performance or political policy. Good fortune and ill lay entirely in human responses to Heaven: "Respond to it with order, and there will be good fortune. Respond to it with disorder, and there will be ill-fortune. . . . To those who practice the way without deviating, Heaven can bring no calamity" 應之以治則吉, 應之以亂則凶 . . . . . . 脩道而不貳, 則天不能禍.[7] A narrative episode in the *Zuozhuan* 左傳 (Zuo Tradition) presented a more subtle instance of this view. In 526 and 525 BCE, two comets appeared, the second of which encountered Great Fire (Da huo 大火), the star Antares. Officials in Zheng 鄭 warned that the comet portended the outbreak of widespread fires in several domains, including their own. They argued that Zheng could be spared by means of a ritual offering of a pouring vessel and a jade ladle. Zichan 子產 (fl. 564–521), the exemplary prime minister, refused to provide these implements. Seemingly unconcerned with the episode, he offered neither an explanation of why the comet had come into being nor a plan for expiating it.[8]

The Eastern Han thinker Wang Chong 王充 (27–ca. 100) presented still another view. Wang saw Heaven as exercising a powerful influence over human beings, providing them with an allotment of *qi* that determined their lifespans, talents, and even career trajectories, but that influence went only one way. Acknowledging that allotments of *qi* were evident in bone structure and that a trained eye could forecast the weather by examining the stars or the clouds, Wang averred that signs existed. Yet, while a ritual expert might know when it was about to rain and dance before it fell, no one could make it rain. In Wang Chong's view, human beings had no capacity to influence Heaven.[9]

Those who sought to make claims about the meaning of omens, including Master Yan and many prominent officials in Han times, positioned those signs in a context. Sometimes that context was the sky itself. Groups of constellations called field allocations (*fenye* 分野) corresponded to terrestrial domains. Interpreters could link an ominous celestial phenomenon in the constellation Barrens (Xu 虛), for example, to political events in the domain of Qi.[10] Master Yan read the comet in conjunction with signs in the human realm: the duke's indulgent behavior, his poor governance, his failure to recognize good men, and the starving and rebellious populace. Like Master Yan, later experts in reading celestial signs understood them as part of the human world rather than natural phenomena entirely separate from it.[11]

Celestial Signs and Shifting Definitions of *Tianwen*

The present study concerns the processes underlying the interpretation and rhetorical appropriation of celestial signs, the claims such signs allowed historical actors to make, and the ways in which those claims influenced and were influenced by other types of knowledge, political interests, ritual concerns, and historical circumstances. As actors at early courts observed, recorded, interpreted, and invoked celestial phenomena, the lives of those courts shaped the meanings of celestial signs. Engaging celestial signs in early imperial China requires coming to terms with the closely related actors' category of *tianwen* 天文.

Tianwen, the modern Chinese word for astronomy, literally means "celestial patterns." Historian of science Jiang Xiaoyuan 江曉原 argues that *tianwen* had two basic definitions in early China. First, it referred to *tianxiang* 天象, celestial bodies or celestial signs. By extension, it also carried the secondary definition of the practical and intellectual field involved in recording and interpreting those signs.[12] In early imperial and medieval times, *tianwen* concerned a different set of phenomena and carried a different purpose than does modern astronomy. While recognizing their considerable overlap in practice, Nathan Sivin treats *tianwen* in premodern contexts as "astrology," reserving the rendering "mathematical astronomy" for *li*:

> In imperial China, astrology (*t'ien-wen* 天文) and mathematical astronomy (*li* 曆, *li-fa* 曆法) were based on the observation and prediction of actual phenomena in the sky. Both were primarily governmental functions, and were interdependent. These complementary arts evolved from the responsibilities of early scribes[13] who kept records and divined to determine whether courses of action were propitious or disastrous.[14]

Li determined regularities, whereas *tianwen* interpreted anomalies. The early sense of *tianwen* was, however, somewhat unstable, and it continued to evolve in meaning even beyond the Eastern Han.[15]

In its earliest contexts, *tianwen* is paired with observations of the human or terrestrial world. In the *Changes* Classic, the "Judgment Tradition" (Tuan zhuan 彖傳) for the hexagram Elegance (Bi 賁) pairs the observation of celestial and human patterns: "Observe celestial patterns so as to investigate the changes of the seasons [or times] (*shi bian*), and observe human patterns so as to transform and bring completion to the world" 觀乎天文, 以察時變; 觀乎人文, 以化成天下.[16] *Tianwen* in this context may have been similar to later *li*, though *shi bian* might also be read "seasonal aberrations" so as to imply omen reading, or "changes of the times" to suggest a search for historical patterns. The "Appended Statements" (Xici 繫辭) to the *Changes* contains the most frequently cited lines on *tianwen*: "The *Changes* is attuned to Heaven and Earth, and so it can be used to accurately describe the Way of Heaven and Earth. [The sages] gazed up to observe the celestial patterns and lowered their heads to observe the contours of the Earth. In doing so, they learned the reasons for obscurity and illumination" 易與天地準, 故能彌綸天地之道。仰以觀於天文, 俯以察於地理, 是故知幽明之故.[17] While each of these statements suggests that much can be gained by observing celestial patterns, neither clearly indicates that *tianwen* is more concerned with reading omens than developing mathematical models or creating calendars.

Treatises from the second century BCE likewise show that *tianwen* did not yet exist in a stable opposition to *li*. The first treatise to include *tianwen* in its title, the "Tianwen" 天文 (or "Tianwen xun" 天文訓)[18] of the *Huainanzi* 淮南子 (Master of Huainan, submitted to the throne ca. 139 BCE) included materials that might be categorized as calendrical or hemerological, in addition to those that seemed to pertain to reading omens. The prototype for later *tianwen* treatises, which are found in most subsequent dynastic histories, did not include the word *tianwen* in its title. Despite its many similarities to the *Hanshu* "Celestial Patterns," the "Tianguan shu" 天官書 in Sima Qian's 司馬遷 (ca. 145–ca. 86) *Shiji* 史記 (Records of the Senior Archivist, ca. 100 BCE)[19] was titled as a "Treatise on Celestial Offices," not celestial patterns.

In the late Western Han, *tianwen* became a bibliographical category focused on the omenological reading of astronomical and meteorological signs.[20] The *Hanshu* 漢書 (History of the Han) bibliographical treatise makes the omenological character of *tianwen* texts clear in its description of them:

天文者, 序二十八宿, 步五星日月, 以紀吉凶之象, 聖王所以參政也。《易》曰:「觀乎天文, 以察時變。」然星事殆悍, 非湛密者弗能由也。夫觀景以譴形, 非明王亦不能服聽也。以不能由之臣, 諫不能聽之王, 此所以兩有患也。

As for *tianwen*, it puts in order the twenty-eight lunar lodges and paces the orbits of the sun, moon, and Five Planets to determine auspicious and inauspicious counterparts. It is the means by which sage-kings engage in governance. The *Changes* states: "He observes celestial patterns in order to examine the changes of the times [*or* seasons]." This being so, when the affairs of the stars surge with ill-fortune, those who are not profound and perspicacious cannot follow them. And thus, as for observing shadows to take stock of the forms that cast them, if one is not a perceptive king, one cannot hope to heed them. When a minister who cannot follow [the movements of celestial bodies and celestial *qi*] advises a king who cannot heed them, both then come to harm.[21]

The subcategory of *tianwen* in the *Hanshu* bibliographic treatise includes works concerning the movements of the sun, moon, and stars such as *Sign-Readings and Verifications Concerning the Procession of the Five Planets, Comets, and Guest Stars in the Han* (Han wuxing huike xingshi zhanyan 漢五星彗客行事占驗). Yet it also includes titles such as *Guo Zhang*[22] *Observations of Rainbows, Clouds, and Rain* (Guo Zhang guan ni yun yu 國章觀霓雲雨), and the *Oceanic*[23] *Account of Various Sign-Readings Concerning the Sun, Moon, Comets, and Rainbows* (Haizhong riyue hui hong zazhan 海中日月彗虹雜占).[24] Appendix B includes a complete list of *tianwen* texts in the *Hanshu* bibliographic treatise.

This bibliographic category aligns with the type of content seen in the *Hanshu* "Celestial Patterns" treatise, and it is this sense of *tianwen* that modern scholars generally have in mind when referring to *tianwen* as an actors' category. However, the category of *tianwen* remained porous, even in the mid-Eastern Han. The *Hanshu* "Celestial Patterns" treatise, like the *Shiji* "Celestial Offices," contains a twenty-four-line poem that includes numerous anomalies and disasters within the terrestrial and human realms. Its opening lines establish the parity between signs in the heavens and on the earth: "Heaven opens and suspends these things / Earth shakes and splits" 天開縣物, 地動坼絕.[25] The poem goes on to describe numerous events that do not seem to fit easily within the category of celestial patterns. Mountains crumble. Marshes dry up. City walls wither away like grass. The poem implores the reader, or perhaps the listener, to be attentive to conditions in the world that allow agriculture and human society to flourish.

The sense of *tianwen* continued to evolve beyond Han times. Treatises on *tianwen* in later dynastic histories were not mere extensions of their *Hanshu* (and *Shiji*) prototypes but added new types of materials that served to expand the sense of *tianwen*. Work by both Ho Peng Yoke and Daniel Morgan has shown that, despite the tendency of *tianwen* treatises to repeat the contents of their predecessors, major innovations in the genre continued into the medieval period, as a new section on astronomical instruments was added by late Six Dynasties historians and further developed by Tang dynasty (618–907) astronomer Li Chunfeng 李淳風 (602–670).[26]

The bibliographic category of *tianwen* is a useful point of departure for a study of celestial signs, as it highlights both the broad scope of phenomena that counted as such signs and the human stakes involved in reading the heavens. Technical manuals included in that category served as guides for interpreting astronomical and meteorological phenomena to understand and prepare for a range of challenges, including military engagements, crop failures, and treachery at court. This book is not, however, a study of *tianwen* in any strict sense. Many texts in the first part of the book were compiled before *tianwen* took shape as a formal bibliographic category in the late Western Han, though the label might reasonably be applied retroactively to some of them. The memorials in the second half of the book did not fall into the *tianwen* bibliographic category but drew on the type of knowledge contained in *tianwen* texts, in conjunction with lessons drawn from the Classics, historiography, and lived experience (among other sources), to interpret both the signs themselves and the contexts in which they appeared.

Celestial Signs across Disciplines

The present volume is a work of cultural history and is reliant on previous scholarship in a variety of fields. Scholars who contribute to our understanding of celestial signs work in multiple modern disciplines: critical theory, literature, art history, archaeology, rhetoric, religious studies, and the history of science. Research on celestial signs has implications beyond the humanities, especially for the emergent field of terra-astronomy.

At the core of this study is a kind of curiosity about the nature of signs, the relationship between a given sign's form and the meaning that is assigned to it, underpinned by well-known theories of the construction and circulation of signs. This book assumes, as Ludwig Wittgenstein (1889–1951) held, that signs have no inherent meaning but mean what people use them to mean. It also assumes, as Julia Kristeva has argued, that meaning can emerge through intertextuality, as instances of words, phrases, and images in one text call to mind other texts where those signs appear. Following the New Historicist tenet that "literary and nonliterary 'texts' circulate inseparably," the analysis in this volume applies a common method to signs that appear in disparate genres.[27] Employing extensive close readings of its sources, this study searches not for the essential meanings of signs but seeks to articulate the processes—hermeneutic, rhetorical, ritual, and historiographical—by which historical actors construct the meaning of signs even as they evoke and interpret them.

Studies in art history and in literature have paved the way for a humanistic approach to celestial signs that includes a broad swath of phenomena, ranging from deities to meteorological events. Edward Schafer's *Pacing the Void: T'ang Approaches to the Stars* (1977) and Lillian Tseng's *Picturing Heaven in Early China* (2011)[28] are two exemplary works. Schafer succeeds in addressing many of the technical, religious, ritual, and literary roles the stars played in Tang culture in part because he did not limit himself to a single, established field of inquiry. Observation of the stars, narratives of descending deities, and poetic production all intermingle. He writes:

> [T]he general trend of my text is far from "systematic" in the best sense of the word. It is based on a selection of ingredients—literary, folkloristic, and astrological—from which I could blend a reasonably rich goulash whose flavor might be distinguished as *a la T'ang* . . . I propose to correct, in some degree, prevailing tendencies to see the T'ang world through European or American eyes—the fallacy of stressing what we stress, of painting with the colors available on our palettes, of fusing only those amalgams which we regard as reasonable or useful.[29]

Employing both received and archaeologically excavated materials, Tseng likewise takes an inclusive approach, treating celestial signs not only as signs appearing in the heavens but as signs of Heaven's favor. In her chapter on "Engraving Auspicious Omens," Tseng examines five signs in the "Hymn of the Western Passage" (Xi xia song 西狹頌): a Yellow Dragon (Huanglong 黃龍), a White Deer (Bailu 白鹿), two trees growing together (mu lian lili 木連力理), a fine sprout (jia he 嘉禾), and the descent of sweet dew (gan lü jiang 甘露降)—all engraved on a cliff side in 171 CE to commemorate the achievements of Li Xi 李翕, governor of Wudu 武都 Commandery.[30] Had Tseng restricted what it means to "picture Heaven" to astronomical phenomena, or even phenomena in the *tianwen* category, these five omens would have remained outside the scope of her study. Tseng shows, however, that in Li Xi's time these omens were signs of celestial favor. In Tseng's work, celestial signs are inscribed into and inseparable from the broader cultural landscape.

Like *Picturing Heaven*, David Pankenier's *Astrology and Cosmology in Early China: Conforming Earth to Heaven* (2013) extensively employs archaeological evidence to produce a richer picture of how the observation of the heavens shaped culture in early China. In contrast to Tseng, however, Pankenier leaves aside transient and meteorological phenomena, focusing instead on problems that benefit from our modern capacity to reconstruct the precise movements of celestial bodies. Central to his investigation is the political significance of a series of planetary conjunctions over some 3,500 years, ranging from 1953 BCE to 1524 CE.[31] Drawing on interdisciplinary scholarship, including the works of religious theorist Mircea Eliade (1907–1986), cognitive scientist Merlin Donald, and anthropologist and archeoastronomer Anthony Aveni, Pankenier offers deep analyses of problems concerning archeoastronomy and culture, tying the observation of the heavens to the construction of prehistorical solar altars, the origin of the graph that came to mean godking or emperor (di 帝), and early imperial architecture.

Celestial signs likewise involve problems of rhetoric. Studies of rhetoric in the *Zuozhuan* address questions concerning how political speeches interpreted, contextualized, and employed various signs, ranging from minute ritual lapses to comets and eclipses. Marc Kalinowski argues that predictions in the *Zuozhuan* served to create a sense of narrative continuity through the whole of the text.[32] David Schaberg's *A Patterned Past* (2001) frames the regularity of the heavens as a model for human morality.[33] Li Wai-yee's *The Readability of the Past in Early Chinese Historiography* (2007) points to the critical importance of reading human ritual lapses—traces and gestures—along with omens that issue from non-human sources. An anomaly is not a meaningful sign, in other words, unless it can be tied to human failure.[34] Kalinowski, Schaberg, and Li all assiduously read the speeches in the *Zuozhuan* as compositions that took their final forms centuries after the speakers had died. The historiographers knew what events had since come to pass, and thus the speakers in the text often attain an uncanny prescience. Han memorials, by contrast, read signs less with respect to events that are certain to occur in the future than with respect to the historical milieux that prevailed at the time of their composition: the personal and ritual comportment of the emperor and his closest companions, the effectiveness of the administration, and the economic and military conditions of the empire. In the present study, questions concerning the construction of authority and the relations among ritual, sign-reading, and classicism are paramount. Core questions include: How did memorialists make claims regarding what signs meant? What kind of epistemology did they employ?[35] What precedents did they evoke, and to what end?

Celestial signs were embedded in both a culture and a world, in social structures and in cosmic ones. Scholarship in the history of science has drawn into focus relations among observation, culture, and world-making. Geoffrey Lloyd and Nathan Sivin, in their comparative study of early Greek and Chinese science, *The Way and the Word* (2002), advanced an imperative to study technical practices within a cultural manifold, a whole that included interrelations and interactions that bind together the "intellectual, social, and institutional dimensions" of technical work.[36] Lloyd and Sivin foreground the relationship between technical work and social circumstances, asking "how people make a living, what their relation is to structures of authority, what bonds connect those who do the same work, how they communicate what they understand, and what concepts and assumptions they use."[37] Celestial signs circulated across textual genres, types of discourse, and social and political structures. Observations and ideas about specific phenomena both fed into and were shaped by broader ideas about the world as a whole. Daryn Lehoux's *What Did the Romans Know* (2012) makes the case well: "The more we think our way into the details, the more tangled and intersected the web in which we find ourselves. We emerge in the middle of a multidimensional world, where politics cannot be cleanly distinguished from conceptions of nature, and neither of them from conceptions of the gods."[38] What writers in the Han empires knew of celestial signs was likewise inseparable from their conceptions of nature, politics, people, and gods—the world as a whole.

This study complements scholarship in the history of Chinese astronomy and astro-omenology but overlaps less with existing works than one might expect. Scholarship on the history of astronomy in Western languages centers on *li* rather than *tianwen*, phenomena that can be understood in precise terms through highly reliable retrospective computation. Eclipses and planetary motions for the past several thousand years can be calculated with precision and accuracy. Scholars working on calculable signs are able to measure historical methods of calculation against modern standards. Sun and Kistemaker's *The Chinese Sky During the Han* (1997) details the meanings of star names and uses mathematical modeling to identify the time when early star catalogues were compiled.[39] Christopher Cullen's *Heavenly Numbers* (2017) shows the development of astronomical models during the Western and Eastern Han, culminating in Liu Hong's 劉洪 (ca. 130–ca. 210) models of lunar motion that allowed for lunar eclipses to be predicted with a greater degree of accuracy than ever before. Daniel Morgan's *Astral Sciences in Early Imperial China* (2017) argues that those who developed increasingly accurate models of planetary, lunar, and solar motion held up "tightness" (*mi* 密) as a cardinal value and that Chinese astronomy progressed gradually over time.[40] Nathan Sivin's *Granting the Seasons* (2009) details the high-water mark of traditional Chinese mathematical astronomy. Completed under the newly established Yuan 元 (1279–1368) dynasty, the *Shou-shi li* (Season-granting system) advanced the legitimacy of the Mongolian rulers by creating the most powerfully predictive system of mathematical astronomy yet developed.

Because modern and premodern senses of *tianwen* overlap, the two are sometimes conflated. Modern scholarship in Chinese concerned with the omenological aspects of *tianwen*, including works by Li Ling 李零 and by Lu Yang 盧央,[41] sometimes instead adopts the terms *xingzhan* 星占 or *zhanxing* 占星, both roughly meaning astro-omenology. In his *Tianxue zhenyuan* (1991), Jiang Xiaoyuan uses the word *tianxue* 天學 (≈celestial studies), citing stark contrasts between both the character and the function of modern and premodern *tianwen*.[42] Historical surveys of *tianwen* in China tend to emphasize the modern sense, focusing on astronomical phenomena rather than meteorological phenomena. Scholarly book-length treatments of *tianwen* written in Chinese, including monographs by Chen Meidong 陳美東 (2008) and by Feng Shi 馮時 (2001),[43] focus on phenomena that fall within the scope of modern astronomy. Chen Meidong writes about models of planetary motion, eclipses, and even the influence of the moon on the tides, but avoids meteorological phenomena that were part of the premodern discourse surrounding *tianwen*, such as rainbows, strange clouds, and halos around the sun and the moon. Feng Shi's writings on archaeoastronomy foreground the religious and ritual aspects of astronomical observation but nonetheless largely eschew meteorological phenomena. This demarcation problem, resulting from the slippage between modern and premodern senses of *tianwen*, can be confusing for outsiders. Modern astronomers using historical records have sometimes assumed that meteorological phenomena were astronomical because they were included in *tianwen* records, negatively impacting their work.

Studies of the full range of celestial signs are critical to the scientific mission of the emerging field of terra-astronomy. Focused on the study of astronomical phenomena that impact the planet Earth, terra-astronomy reconstructs past solar activity and explosions of nearby stars using terrestrial archives. In addition to physical archives such ice cores, tree rings, and radioisotopes, these include written records of celestial phenomena. Solar and lunar halos, caused by the presence of ice crystals in the upper atmosphere, are useful for reconstructing weather. Aurorae, caused by the impact of charged particles from the sun with Earth's magnetic field, are a critical source for astrophysicists working to reconstruct past solar activity.[44] Readers who do not recognize the presence of both meteorological and astronomical components of East Asian records of celestial signs confuse records of aberrant weather and atmospheric optical phenomena for astronomical events. This sometimes leads to serious errors.[45] Failure to consistently keep in mind the full scope of the celestial sciences in premodern East Asia is damaging to current scientific efforts to understand solar cycles and other phenomena that have a measurable impact on Earth.

Audience, Limitations, and a Word on Teleology

This book is aimed at two distinct audiences: (1) readers with sinological backgrounds specifically interested in early China, and (2) readers interested in celestial signs who have no prior background in early China.

Such readers might include historians, scholars of religion, or cultural astronomers working on South Asia, the Near East, West Africa, Mesoamerica, or the Mediterranean World, among other regions. The book includes historical background that readers who know early China well likely will not need, and it presents summaries of information culled from the works of historians of science, especially in its first chapter, that is well established within the field. Most discussion of previous scholarship is confined to the notes, except where it is quite necessary to advance the main arguments of the book.

While this study examines a broad range of signs in Han times,[46] working to show connections among genres, it does not go as deep into any particular genre as a more specialized volume might. A full monograph might be written on technical literature at Mawangdui, parallel passages in the *Shiji* and *Hanshu*, or sign-reading in Han memorials. While I hope that the volume will be of some use to those pursuing comparative scholarship, the book itself has no pretensions to being a comparative work. Even its engagement with celestial signs in Han times is limited to a few particular strands of development. It only briefly touches on the apocrypha, as these materials survive only in fragments and do not seem to have been produced by any of the key figures that the book follows. It leaves aside important early uses of celestial signs, such as the diviner's boards that placed the Dipper at the center and the twenty-eight lunar lodges at the periphery, as well as the use of celestial signs in excavated hemerological manuscripts, where the twenty-eight lodges appear as a kind of day-count.[47] This is in part due to the volume of excellent scholarship already done in this area, and in part because this type of material shows more concern with regularities than with sign-reading. Another significant omission is an investigation into the tradition of celestial journey poetry, liturgical in origin, but increasingly literary over time. Early liturgical texts represent acts of supplication to deities that appear in texts for sign-reading at Mawangdui, and several figures who loom large in this volume, as both technical experts and memorialists, produced literary representations of celestial journeys with classical overtones. A study of celestial journey verse involves methods and materials that are too far afield to include in this volume, but nonetheless represents a promising research direction. Indeed, with regard to figures such as Liu Xiang 劉向 (79–8 BCE), Zhang Heng 張衡 (78–139), and Ma Rong 馬融 (79–166), it is worthwhile to pursue a fundamental question posed by Howard Goodman regarding polymaths: "what else was he a master of and why"?[48]

While it has been necessary to leave aside certain materials to produce a monograph with a clear argument and manageable length, doing so comes with a certain risk that is best addressed in an explicit fashion. One possible misreading of this volume is teleological: "Technical materials and the Classics existed in the early Western Han and *inexorably* became bound up with one another." The most obvious problem with this comes in the word *inexorably*, as though no other outcomes were possible. A more subtle problem, however, is that not all technical materials became bound up with the Classics, which were themselves still in the process of formation. A study examining celestial signs across genres in the early Western Han, without respect to what came later, would certainly include materials not examined in this volume. I aim to show developments, but not to claim (or even suggest) that those developments were inevitable.

The Heart of the Argument

The central argument of this book consists in two interlocking strands, the first concerning the relationship between technical and classical discourse in Han times, and the second concerning the process of observing, interpreting, and invoking celestial signs on a more general level.

First, the book traces the gradual interweaving that occurred between two types of discourse: nitty-gritty omenological knowledge found in technical manuals, and theories, narratives, and rhetorical structures concerning signs and sign-reading found in classical texts. At the outset of the dynasty, classical texts and technical texts could be found in the same communities, but they did not interpenetrate on the textual level. The same people sometimes read and used such texts, and the presence of classical theories concerning the obligations of the ruler to his subjects, for example, might have influenced how people in those communities understood sign-reading texts that centered on protecting the ruler's life and maintaining his power. Later in the second century BCE, large-scale textual projects, including Liu An's 劉安 (179–122) *Huainanzi* and Sima Qian's *Shiji*, integrated technical sign-reading texts into broader ideas concerning statecraft. Sima Qian treated the reading of celestial signs as an integral component of sagely governance, describing the greatest rulers as those most gifted in interpreting signs. As a new form of classicism emerged in the late Western Han, celestial signs were integrated into systematic treatises that emulated classical texts and drew directly on classical authority. Claiming that celestial signs meant the court should take certain actions, memorialists called on precedents and etiological theories drawn from the Five Classics and other classical texts.

Second, the book addresses the problem of how celestial signs fit into world-making. It claims that sign-reading in practice proved both flexible and context dependent, notwithstanding the seemingly systematic nature of technical texts on omen interpretation. Interpreters of celestial signs rarely, if ever, read those signs in isolation. Much like erudite interpreters of texts, they read signs against the context of their appearance. Their memorials responded not just to the signs themselves but to conditions at court, in the palace, and throughout the empire, as well as to past appearances of similar signs and the events that followed them. Just as readers of words in classical texts understood each word as part of a larger context, interpreters of celestial signs read each sign as part of a larger world.

Outline of Chapters

Part I, "Technical Manuals and Classical Authority," focuses on the development of the technical discipline of *tianwen* and the growing role classical authority came to play in it in Han times. The first chapter, "Practical Omenology and Classical Ideals at Mawangdui" examines a corpus of texts interred early in the Western Han dynasty (168 BCE). It shows that while classical texts and technical sign-reading texts could be found within a single community, technical manuals did not yet explicitly cite or refer to classical authorities. The second chapter, "Sign-Reading and Sagely Rulership in the *Huainanzi* and *Shiji*," turns to the first large-scale textual projects to include systematic treatises on *tianwen*, dating respectively to ca. 139 BCE and ca. 100 BCE. These textual projects show that technical knowledge had begun to be integrated with narratives surrounding exemplary figures from the classical tradition, particularly in the case of the *Shiji*. They represent a transitional phase from a pragmatic and largely amoral tradition of reading the heavens to a technical practice anchored in classical ethical ideals. The third chapter,

"Integrating Classical Texts with Omen Reading Technology," turns to two treatises included in the *Hanshu*, the "Wuxing Treatise" (Wuxing zhi 五行志) and the "Treatise on Celestial Patterns" (Tianwen zhi 天文志). The chapter shows that these treatises fully integrated the technical practice of sign-reading with classical ethical ideals. The "Wuxing Treatise" drew its structure from the *Annals* Classic and the "Great Plan" (Hong fan 洪/鴻範) chapter of the *Documents* Classic. The "Treatise on Celestial Patterns," though largely drawn verbatim from its earlier counterpart in the *Shiji*, introduces new material that frequently cites the Five Classics as proof-texts, relies on ethical ideals from the "Great Plan," and mirrors the structure of the *Annals* Classic. The treatises show that by the early Eastern Han, the technical practice of reading celestial signs had become inseparable from classical authority.

Part II, "Rhetoric in Practice," turns from technical texts to memorials, engaging with the practice of reading celestial signs in court rhetoric.[49] Chapter 4, "Classical Authority and the Elimination of Baleful Signs," shows that memorialists from the late Western Han on routinely cited the "Appended Statements" section of the *Changes* Classic, the "Great Plan," and a set of precedents concerning classical exemplars who responded well to the appearance of baleful signs. Diverse as their individual aims may have been, skilled memorialists deployed classical authority to argue that the sources of baleful signs could be identified, and through the correct course of political, ritual, or military action, both the signs themselves and the unfortunate events they presaged could be eliminated. Chapter 5, "Reading Signs, Reading the World," closely analyzes three case studies: the absence of auspicious signs early in the reign of Emperor Wu 武 (r. 141–87), the appearance of Halley's Comet in 12 BCE, and the earthquake in the capital in 133 CE. Though technical texts created a sense of systematic rigor surrounding omen interpretation, this final chapter demonstrates that the practice of reading signs remained subject to individual aims and methods. Memorialists read signs against the state of the court, the state of the empire, and other signs in both the distant and recent past. To read any given sign was to read the state of the world at a particular moment in time.

Part I
Technical Manuals and Classical Authority

Chapter 1

Practical Omenology and Classical Ideals at Mawangdui

Early manuscripts concerned with sign-reading seem at first blush to fall into disparate, almost incommensurable categories. At one end of the spectrum are manuals concerned with practical omenology, bare lists of signs that present technical details necessary for identification and interpretation. At the other are works concerned with broad questions of governance or cosmology, engaging what might be called the theory of sign-reading. These include writings that call for modeling human society on the regularity of the heavens, as well as texts that explain how sign-reading works and why it often fails. Emphasizing the bonds between sign-reading, character, and good governance, these texts are loosely representative of classical ideals concerning omenology—though only a fraction of them were ultimately preserved in the Classics.

The first three chapters of this volume are largely the story of how practical omenology and classical ideals became increasingly intertwined and interdependent, to the point that, by the Eastern Han, treatises on sign-reading routinely integrated quotations from the Classics as prooftexts. In the early Western Han, this process remained at an inchoate stage. Practical manuals for reading celestial signs had yet to be textually interwoven with works that directly addressed the ethical aspects and cosmological theory of sign-reading. Yet conditions existed that would allow such an interweaving to occur, as these seemingly disparate texts were collected in common textual communities.

One such community was the Li 利 clan at the court of the Kingdom of Changsha 長沙, whose textual corpus was interred in the third tomb at the Mawangdui 馬王堆 archaeological site. It is this community, rather than the individual tomb occupant, that best serves as a lens for considering the ways in which the texts found in the tomb relate to human concerns. Among the corpus are a number of technical manuals for reading celestial signs, as well as texts concerned with the political, ethical, and cosmological aspects of sign-reading in general. This chapter situates the texts both in relation to one another, and in relation to the historical circumstances in which members of the textual community lived. Following an overview of their archaeological and historical context, and a brief discussion of the methodological problems of interpreting fragmentary excavated texts, close readings of three technical manuscripts will examine a range of signs and interpretations to elucidate the most pressing problems in the texts: agricultural production, warfare, and maintaining political power. These will be followed with close readings of passages from other texts in the same tomb that frame the observation and emulation of the heavens as a means of creating a stable, balanced order in the human realm, while casting sign-reading as a fraught, error-prone process largely dependent on the character of the sign-reader.

The Tombs of the Marquises of Dai

When Li Cang 利蒼 (fl. 193–186), Chancellor of the Kingdom of Changsha,[1] was ennobled as the Marquis of Dai in 193 BCE,[2] the notion of a unified empire was new and the future of the dynasty uncertain. The preceding three centuries, the Warring States period (481–222), had been a time of near constant warfare between rival polities. Powerful states collapsed, ministers and generals usurped power, and in 256 BCE the state of Qin deposed the last king of the Zhou (1046–256) dynasty. In the late 230s and 220s, the Qin had rapidly conquered what remained of the old powers, establishing the first empire in 221. Rebellion and civil war broke out following the First Emperor's death in 210.[3] While Liu Bang 劉邦 (r. 202–195) officially founded the Han dynasty in 202, he continued to fight to consolidate the empire until his death from an arrow wound seven years later.[4] The Western Han itself almost fell in 180, when clansmen of the dynasty's first empress[5] attempted to seize power.

In 168, the Li clan buried an honored kinsman, most likely identifiable as the second Marquis, Li Xi 利豨 (r. 185–168 or 164).[6] The clan would maintain possession of Dai, a district of some seven hundred households from which they were entitled to collect taxes, until the fourth marquis was stripped of the nobility for providing an unauthorized military escort during his tenure as governor of Donghai 東海 commandery in 110.[7] Little is known of the tomb occupant's life. It is likely that he, like Li Cang, served in some capacity at the regional court in Changsha. He appears to have died in his thirties, and given the riches of his tomb, he no doubt enjoyed wealth and high status. His kinsmen lavishly appointed his tomb with three nested coffins, hundreds of luxury items, and a T-shaped banner depicting a celestial journey. Between the second and third coffins, they placed many of the same types of objects the tomb occupant might have used in the living world—a sword, two bows, several musical instruments, a fine set of ivory utensils, and a game akin to a chess set. They buried doll-sized statues of servants and entertainers, along with miniaturized replicas of bronze bells and chimes.

Members of the Li clan likewise interred a corpus of writings that would have been useful to any powerful political family in uncertain times.[8] They included treatises on effective governance such as the otherwise lost *Constant Models* (Jingfa 經法), a text inscribed on the same manuscript as a copy of the *Daodejing* 道德經. They included texts on virtuous action, such as the *Five Kinds of Action* (Wuxing 五行). They included texts on sign-reading and the theories surrounding it, such as the *Changes* Classic (Zhouyi 周易), as well as both received and previously unknown traditions attached to it. They included medical texts that provided for the health of the ruling family, and critically, ensured that the clan would continue to conceive and bring to term heirs.[9] Along with numerous other texts, the clan placed in the tomb a series of technical manuscripts for interpreting various types of celestial signs, ranging from eclipses and planets in retrograde motion to the appearance of oddly shaped clouds, comets, and rainbows. The manuscripts identified these signs as harbingers of military victories[10] and defeats, bumper crops and famines, the rise and fall of kings.

Situating the Significance of Excavated Texts in Textual Communities

Both the physical condition of excavated texts and the context of their use and transmission make them difficult reading for modern scholars.

Even in the ancient world, manuscripts themselves were sometimes inconsistent, and reading them often required engagement with a teacher or prior knowledge of the text.[11] To understand why a given text was significant enough to be placed in the tomb, two related questions must be addressed: What does the text mean (in terms of its semantic, denotative sense)? Why does the text matter? These two questions present distinct challenges.

Coming to terms with semantic meaning of the language of excavated texts involves resolving a number of problems. Excavated texts are almost always damaged, and therefore incomplete. They use nonstandard, variable orthography. Experts in epigraphy compare graphs that are difficult to identify with other instances in the same manuscript or set of manuscripts, consult dictionaries of orthographic variants, and use reconstructions of ancient pronunciation to piece together modern transcriptions of ancient texts. Ideally, scholars rely primarily on evidence present in the manuscripts themselves before looking to outside sources to identify particular graphs.[12] In many cases, however, epigraphers do fill in missing sections of manuscripts based on analogy or closely corresponding sections in received texts—a practice that tends to homogenize manuscript sources to the received tradition.[13] Half a century after the initial discovery of the texts at Mawangdui, annotated transcriptions of the entire corpus are now available.[14]

The question of why texts matter may be addressed only with respect to a given audience or community. Asking why a text matters inevitably raises the question, "Why does it matter to whom?" Thinking through this question disambiguates modern communities from ancient ones. Early studies of the Mawangdui celestial signs manuscripts by Xu Zhentao 徐振韜 (1936–2004),[15] Xi Zezong 席澤宗 (1927–2008), and Gu Tiefu 顧鐵符 (1908–1990) declared the scientific value of the texts,[16] their significance for locating a deeply rooted inclination toward scientific thinking among an ancient people that could serve to inspire their modern descendants. Gu Tiefu urged scholars to leave the "superstitious" aspects of the manuscripts buried, arguing that investigating such materials could only bring harm to the modern nation-state.

Reading a group of texts in historical context involves interpreting their content relative to the concerns of a community that possessed, employed, or compiled those texts.[17] While it is perhaps impossible to leave modern concerns aside entirely, historically situating an excavated corpus involves reading texts with respect to what is known of those who buried them or were buried with them and the circumstances of the decades preceding their interment, as well as thinking through the thematic relationships between the texts. In this case, those who buried the manuscripts belonged to a clan of nobles in the early Western Han, at least some of whom served in high-level positions in regional government; the texts may also be partially representative of a broader textual community at the Changsha court, where they were likely compiled.[18] Drawing out the thematic relationships between the texts is a difficult task, but a necessary one, if we are to read a corpus employed by an ancient community rather than a series of isolated documents neatly cordoned off by modern specialists. Nonetheless, attempting to read a corpus in terms of why it might have mattered to a given community can produce only tentative results. Scholars can read texts with respect to the historical circumstances surrounding their consumption and production, identifying their shared themes and motifs, but cannot see inside the minds of the people who used them. This chapter works to contextualize the celestial signs manuscripts by situating them with respect to both a common community and other manuscripts in the same tomb.

The Celestial Signs Manuscripts

Three technical manuscripts in the tomb center on the reading of celestial signs. Modern epigraphers have provided these with descriptive titles: the "Five Planets Sign-Readings" (Wuxing zhan 五星占), "Harshness and Favor" (Xingde 刑德), and "Miscellaneous Sign-Readings on Astronomical and Meteorological Phenomena" (Tianwen qixiang zazhan 天文氣象雜占). The tomb includes multiple copies of "Harshness and Favor." Following a brief discussion of the organization and physical features of these texts, we will consider their omenological content and concerns:

(1) The "Five Planets," 221 × 48 cm, on silk.[19] The bulk of the somewhat tattered manuscript consists of sign-readings for the retrograde motion, apparent color, or scintillation of each of the five visible planets. The concluding section presents a precise account of the daily motions of Saturn, Jupiter, and Venus, and retrospectively charts the first apparent rising for each over a seventy-year period.[20] This portion of the manuscript has sometimes been treated under a separate title, the "Five Planets Table of Motion in *Du* (≈Degrees)" (Wuxing xingdu biao 五星行度表). An excellent technical translation by Christopher Cullen is available, and both Cullen and Daniel Morgan have produced studies of its mathematical and calendrical aspects.[21]

(2) "Harshness and Favor," A and B, both on silk.[22] "Harshness and Favor B," the most complete of the manuscripts, as well as the basis for Marc Kalinowski's study of the text, measures 84 × 44 cm. Both manuscripts contain a Nine Palaces diagram (Jiugong tu 九宮圖), a sexagenary grid, a meteoromantic section of text, and a section of text pertaining directly to the annual cycles of Harshness and Favor.[23] The meteoromantic sections of these manuscripts are sometimes treated as separate texts under the title "Ri yue feng yu yunqi zhan" 日月風雨雲氣占 (Sign-Readings on the Sun, Moon, Wind, Rain, and Cloud Qi).[24]

(3) The "Miscellaneous Sign-Readings," 150 × 48 cm, on silk.[25] The "Miscellaneous Sign-Readings" is sometimes described as if it consisted only in a chart of comets,[26] but it in fact includes images of clouds, rainbows, eclipses, and halos, coupled with sign-reading statements. Like the "Five Planets," the manuscript is somewhat damaged, but many portions of it, including twenty-nine entries on comets, survive intact. The six horizontal ranks of its main body each contain between twenty-two and fifty-two distinct entries. The first rank consists primarily of images of clouds, while the second through fifth ranks consist primarily of images of various sorts of halos, often in conjunction with the sun or the moon. The third and fourth ranks are badly damaged. Images of rainbows, comets, and a Dipper-shaped cloud occur in the sixth rank. The end of the chart contains sign-reading statements without images.[27]

THE FIVE PLANETS

"The Five Planets" is the best known of the celestial signs manuscripts from the Mawangdui tombs. Its astronomical features and significance in the history of astronomy are well established.[28] A close reading of the text, focused on its omenological features, will serve both to develop our understanding of the text's role in the lives of its users and to establish a basis for investigating its relationship with other texts in the tomb.

The bulk of the manuscript works to identify the omenological significance of various phenomena surrounding each of the Five Planets, proceeding through Jupiter, Mars, Saturn, Mercury, and Venus. The concluding portion of the text lists the position among the stars of three of those planets, Jupiter, Saturn, and Venus, between the year Ying Zheng 嬴正—the future Qin Shihuang 秦始皇 (r. 221–210)—became King of Qin (246 BCE) and 177 BCE. It also gives highly precise figures describing the movements of these three planets over time. The positions of the planets are based on retrospective calculations, and the values given for their movements are produced through dividing larger, rounded wholes.

Each entry for a planet in the "Five Planets" begins by giving a direction and a material resource, followed by the possessive particle *qi* 其, and the name of a god, the god's assistant, and finally a spirit. The spirit is, acts as, or becomes (*wei* 為) the planet itself. The opening for Venus is typical: "The West corresponds to metal. Its God is Shao Hao. His Assistant is Ru Shou. Its spirit ascends to become Great White" 西方金, 其帝少浩 (皞), 其丞蓐收, 其神上為太白.[29] The direction, rather than the planet itself, is the initial subject of the taxonomy. The spirit associated with that direction manifests in the appearance and movements of the planet. With the exception of Jupiter, each of the planets includes a second taxonomic passage near the end of its section.[30] Here, too, the entry for Venus is typical of the second set of correspondences: "Its season is autumn. Its days are *geng* and *xin*. Its lunar position is in the west.[31] The western domains possess it. It is the Master of Celestial Sacrifices" 其時秋, 其日庚辛, 月立 (位) 失 (昳), 西方國有之。司天獻.[32] Each of the Five Planets is embedded in such a taxonomy (see table 1.1), associated with directions, materials, times, deities, and mastery over some celestial domain: prisons, music, rites, virtue (*de*), or sacrifice.

Jupiter and Saturn, the Year Star (Sui xing 歲星) and the Quelling Star (Zhen xing 鎮星), are the most auspicious of the planets. It takes a long time for each planet to complete its journey through the stars, nearly twelve years for Jupiter and nearly thirty for Saturn. They proceed through the constellations in a slow and relatively predictable fashion. Their presence in a celestial field means good fortune for the corresponding domain. When Jupiter resides in the celestial field corresponding to a domain, it piles up grain if it possesses virtue, weapons and armor if it does not.[34] The presence of Saturn provides protection. The text warns those who would attack a domain where Saturn dwells that they risk both their own lives and the futures of their descendants.[35] Where Saturn goes, domains enjoy good fortune and acquire land. If it dwells in the field corresponding to the same domain for a long period, "the domain will be possessed of virtue and land: good fortune" 其國有德、土地, 吉.[36] The inverse is also true. The unexpected absence of these two planets means bad fortune. When Saturn leaves a domain, invasion and ill-fortune often follow.[37] The text associates a range of disasters with Jupiter moving faster or slower than predicted: "In such years, there will be great flooding in the world, or if not, the sky will rend, or if not,

Table 1.1. Planetary Correspondences in the Mawangdui "Five Planets"

	Jupiter	Mars	Saturn	Venus	Mercury
Direction	East	South	Center	West	North
Material	Wood	Fire	Soil	Metal	Water
God	Tai Hao 太皞[皞]	Red God 赤帝	Yellow God 黄帝	Shao Hao 少浩[皞]	Zhuan Xu 端玉[顓頊]
Assistant 丞	Goumang 句芒	Zhu Ming 朱明	Lord Soil 后土	Ru Shou 蓐收	Xuan Ming 玄冥
Spirit	Year Star 歲星	The Dazzling Deluder 熒惑	Quelling Star 鎮星	Great White 太白	The Watch Star 晨[辰]星
Season	Spring	Summer	The Center	Autumn	Winter
Days	jia, yi 1–2	bing, ding 3–4	wu, ji 5–6	geng, xin 7–8	ren, gui 9–10
Lunar Position	East 東方	Near the Zenith 隅中	Exact Zenith 正中	Moving West 失[昳]	West 西方
Direction	The Eastern Domains 東方之國有之	The Southern Domains 南方之國有之	The Central Domains 中國有之	The Western Domains 西方國有之	The Northern Domains 北方國有之
Celestial Master	Master of Celestial Prisons 司天獄	Master of Celestial Music 司天樂	Master of Celestial Rites 司天禮	Master of Celestial Sacrifice 司天獻	Master of Celestial Virtue 司天德

there will be earthquakes" 於是歲天下大水, 不乃天死〈列 (裂)〉, 不乃地動.[38] The text associates aberrant planetary movements with floods and earthquakes, disasters in the world below that memorialists at later courts would interpret as celestial signs in themselves.[39] When the movements of Jupiter became erratic, it spelled disaster for both the corresponding state and the whole of the known world.

Sections on Venus and Mercury likewise show the bond between planetary anomalies and sublunar phenomena. Because they are closer to the sun than Earth is, both planets always appear relatively close to the sun. Venus is never more than 47 degrees from the sun, Mercury never more than 28. This proximity colors certain signs associated with the two planets. The text posits that it is constant or regular (*jing* 經) for Mercury to appear at the solstices and equinoxes,[40] but because it comes into conjunction with the sun approximately every 116 days, no such regularity exists in nature. The text explains that if Mercury appears earlier than expected, lunar eclipses occur, and if later, comets and celestial monstrosities (*tian yao* 天夭) do.[41] The absence of Mercury for four seasons means "great famine in the world" 天下大饑.[42] Venus, the brightest object in the sky after the sun and the moon, often appears as the first star in the evening or at dawn. The "Five Planets" refers to it as Favor (*De* 德) when it appears in the eastern morning sky and as Harshness (*Xing* 刑) when it follows the setting sun in the west.[43] Both planets resonate with the broader cosmos. Earthquakes occur if Mercury appears between the lodges Chamber and Heart.[44] Venus holds sway over celestial, terrestrial, and human phenomena: "It commands the movements of the moon, broom-stars, celestial monstrosities, armor and weapons, floods and droughts, death and mourning" 是司月行、槍[45] (彗) 星、失〈天〉夭(妖)、甲兵、水旱、死喪.[46] The planets participate in a broader cosmos and the human world, sometimes as harbingers of good fortune, sometimes as the vanguard of troubled times.

The presence of Mars, the Dazzling Deluder (Ying huo 熒惑), in particular portends war and disaster. Unanticipated episodes of retrograde motion are deeply disconcerting: "When it appears in the east, and reverses its course moving back by one lodge, (the domain) from which it has departed will enjoy good fortune, while the domain to which it moves will suffer military incursions" 其出東方, 反行一舍, 所去者吉, 所之國受兵.[47] This portends good fortune for the domain corresponding to the location where Mars ought to be, and bad for the domain corresponding to the location where Mars actually is. It is especially inauspicious for any domain into which Mars might loop-the-loop back, and worse still if it twinkles and reddens.[48] Any deviations from the normative path indicate disaster for the domain corresponding to the celestial field where they occur: "When the Dazzling Deluder strays from the path, the lot of the domain corresponding to its (celestial) field [will suffer calamities]" 營(熒)或(惑)絕道, 其國分當其野【受殃】.[49] In a time of great political uncertainty, the "Five Planets" constructed the Dazzling Deluder as a powerful sign that mirrored the instability of the political order. Should

Mars ever rise from the western horizon, this would portend far-reaching consequences for the empire, regional rulers, and official families: "When it emerges from the west, this is called 'The Reversal of Illumination.' The world changes its King" 其出西方, 是胃 (謂) 反明 (明), 天下革王.[50] For older generations, this passage might have provoked powerful, recent memories: first, the Qin conquest of much of the known world, and then, the dynasty's precipitous fall and the rise of the Han.

Though the body of the manuscript is organized into sections for each of the five visible planets, it includes materials that do not fit in easily, suggesting it was compiled from multiple source texts.[51] Near the conclusion of the section on Jupiter, several entries for comets have no clear relationship to the planet. At least two are weapons: a celestial spear (*tian qiang* 天鎗【槍】), associated with the deity Buzhou 不周, who governs killing, and a celestial staff (*tian bang* 天㭊【棓】) that governs thunder and earthquakes.[52] The section on Jupiter also contains passages that interpret the meaning of the moon occulting the various planets and the bright star Big Horn (Da jiao 大角, Arcturus).[53] The best evidence for the composite nature of the text is perhaps the multiple names it uses for various planets. One part of the section on Venus uses the variant names Yin 殷 and Xiang 相 for Venus and Jupiter.[54] The section on Mercury calls the planet the Watch Star (Chen xing 晨/辰星),[55] but the section on Venus refers to Mercury as Lesser White (Xiao bo 小白).[56]

The subject of extensive studies by both Morgan and Cullen,[57] the group of tables that comes after the main body of the text serves to establish regularities. Text proximate to the tables gives precise figures for the daily movements of Jupiter, Saturn, and Venus, corresponding to cycles of exactly twelve and thirty years for Jupiter and Saturn, and exactly five cycles every eight years for Venus, or one cycle every 584.4 days. These values are fairly close to modern figures for the sidereal periods of Jupiter and Saturn, 11.86 and 29.46 years, and quite close to the synodic period of Venus, 583.92 days. The tables provide the constellations in which the apparent morning risings for Jupiter and Saturn ostensibly occurred between 246 and 177, establishing normative patterns for the planetary movements, but this information is calculated rather than observed. Cullen shows that for the seventy years recorded for Saturn, the given lodge is accurate only thirty-four times. For Jupiter, it is correct only twenty-four times. The table for Venus is more complex, including first and last morning risings and first and last evening settings, but no more accurate. At most, it records the actual lodge seven times in twenty over the course of an eight-year cycle.[58] Recalling the work of David Brown on planetary omens in Mesopotamia, Cullen writes: "One of the advantages of schematic depictions of celestial motions is that they automatically generate portents through their divergence from what is actually observed."[59]

Should regional rulers have found themselves contemplating violence against a neighboring region corresponding to an adjacent lodge, their advisers might have consulted the "Five Planets." Officials versed in sign-reading might have appealed to the current state of the heavens to encourage the ruler to adopt an aggressive stance or a conciliatory one, to go to battle or to engage in diplomacy. The text forged linkages between signs in the heavens and the world below, associating planetary aberrations with political disorder, earthquakes, and devastating weather events—floods and droughts. Yet, the absence of a causal link between the activities of the court *prior to* the appearance of a given sign and the sign itself may have limited the text's rhetorical power in arguments for reforms to political policies or ritual practices. A technical expert could say only what course should be taken *after* the sign appeared, instructing the ruler to make certain sacrifices, or to avoid going to battle. In contrast to technical treatises developed later in the Western Han and in the Eastern Han, the subjects of chapters 2 and 3 in this volume, the "Five Planets" provided little in the way of rhetorical tools to authorize a claim that the ruler had caused the signs to appear in the first place, or that practices already underway prior to the appearance of the sign should be abandoned or reformed.

THE MISCELLANEOUS SIGN-READINGS

The "Miscellaneous Sign-Readings" hints at a historical consciousness centered on Chu, the large, powerful Warring States domain where the Kingdom of Changsha was located, and it concerns many of the same problems that the "Five Planets" does. The constellation of harvest, hunger, funerals, war, and weather runs through the text, along with concerns regarding dangerous ministers, the fate of particular domains, and large-scale political and military developments.

The manuscript makes direct historical references in only a handful of entries. Its regional perspective appears in two entries that center on Chu. Its first rank includes a series of clouds corresponding to different polities in the Warring States world, with Chu first among them. Most take the form of animals, but the Cloud of Chu is represented as a bright sun.⁶⁰ Another entry associates a sun surrounded by three halos with an attack on the Chu capital of Ying carried out by the neighboring domain of Wu 吳. The text likely referred to an attack that had taken place some three centuries earlier, in 506 BCE.⁶¹ The manuscript also evinces an awareness of stories surrounding the fall of the Shang dynasty (ca. 1600–1046 BCE) and the fall of the mythical last ruler of the Xia dynasty (trad. ca. 1600), Jie 桀. Occurring at the end of a long sequence of signs concerning the moon, various entries point to the collapse of the two dynasties. The manuscript attributes these entries to two traditions, one bearing the name of a certain Ren shi 任氏, the other labeled as Bei Gong 北宮 (Northern Palace).⁶²

Many more entries, however, express an immediate sense of instability in the political environment, the court, or the natural world. The manuscript fits into an unstable, chaotic world, associating certain signs with widespread political fragmentation, regicide, and hunger. A caption reading "the world loses its master, and splits into thirty" 天下亡其主，分卅⁶³ appears below an image of a red sun flanked on three sides. Another entry points to the Dazzling Deluder as a harbinger of deceit and strife: "Heaven brings forth the Dazzling Deluder. Those in the world delude one another, bringing forth all of their weapons and armor" 天出熒 (熒) 或 (惑)，天下相惑，甲兵盡出.⁶⁴ One particularly powerful entry, paired with an image of clouds obscuring the sun, presents two possible readings, both of which were available to users of the text. Parsing *jiang* 將 as "general" it reads, "The ministers and generals assassinate the ruler." *Jiang* can also be read as a modal particle, explicitly marking the incipience of the action of assassination: "The ministers are *about to* assassinate the ruler"⁶⁵ 臣將式 (弒) 其君. Pairs of rings shown beneath an image of a rainbow likewise suggest that the ruler's own men are plotting against him. One such image indicates "ministers and lords exchanging position" 臣主貿処 (處) while another points to "ministers attacking their lord" 臣攻主.⁶⁶

Weather events and hunger feature prominently in the text. An entry on the "Plume Star" 翟星 associates its appearance at different times of year with different types of calamitous weather events that would undoubtedly affect the food supply: "If it appears in summer, there is drought. If it appears in fall, there is flooding" 夏見 (現) 旱，秋見 (現) 水.⁶⁷ Certain types of comets indicate not only a hungry populace but also unfed armies. A caption paired with a double-tailed comet reads: "These are mugwort broom-stars. War arises, and the troops go hungry" 是㊀ (是謂) 蒿彗, 兵起, 軍饑.⁶⁸ A three-tailed comet upsets the familiar relationship between famine and war. Its caption reads: "These are sweeping broom-stars. There is civil war (*nei bing*), and that year, a great harvest" 是㊀ (是謂) 帚彗, 有內兵, 年大孰 (熟).⁶⁹

The manuscript likewise associates certain signs with the establishment of new regimes that emerge from disorder. The opening of the fourth rank shows a series of suns surrounded by unbroken halos, which correspond to the establishment of a new political order. With the first, "the Son of Heaven is established" 天子立. The next several correspond to the establishment of noble supporters and regional kings. "There are elders" 又 (有) 柏 (伯) 者. "There are kings" 又 (有) 王者. The fourth image, which portrays a bright streak directed at the sun, might correspond to the establishment of interstate relations. "Within a month, an emissary will arrive" 不出一月, 又 (有) 使至.⁷⁰ The experts who consulted the manuscript lived in a world where the political order had proved unstable and regime change was an ever-present possibility.

At the time the tomb was sealed, the establishment of both the Qin and Western Han dynasties remained in living memory. Older members of the regional court at Changsha had lived through the cataclysmic political changes of the last decades of the Warring States period, the Qin unification and collapse, and the internecine wars that had continued into the early years of the Han empire. While the manuscript may not have been compiled with these events in mind, they made up the fabric of the world older generations had experienced. In the absence of written exegesis or other supplementary materials, it is a fraught task to posit a specific, concrete interpretation for how the community might have received a given manuscript. But whatever range of readings of the manuscript existed, these no doubt derived their significance from the correspondences users saw between it, other texts, and their experiences of the world.

The Fog of War: "Harshness and Favor"

Many celestial signs are meteorological, including solar and lunar halos, oddly shaped clouds, unseasonable weather, rainbows, hazes, and mirages. Early technical texts generally describe these as formations of *qi*. The "Harshness and Favor" manuscripts concern the interpretation of such signs, often reading them in conjunction with the cosmological correspondences associated with the day of their occurrence. These signs frequently presage military conflict but do not always clearly indicate victory or defeat.

"Harshness and Favor" makes predictions based on the appearance of certain kinds of weather at particular times. Rain and lightning on certain days indicate that troops in the field will engage in battle within a specific period of time. Inclement weather on day 1 of the sixty-day cycle means there will be battle within a hundred days, for instance.[71] Fierce summer winds indicate various outcomes based on the month of their occurrence. Winds in the fourth month mean war in fall, while winds in the fifth and sixth months respectively indicate war in the winter and spring.[72] Wind and rain operate primarily as portents of war.

The manuscripts likewise associate military and agricultural sign-readings with deities, weather events, and emissions of qi. The gods Lord Thunder (Lei Gong 雷公), Feng Long 豐隆, Earl of Wind (Feng Bo 風伯), Master of Rain (Yu Shi 雨師), and Great Sound (Da Yin 大音) are tied to two of the twelve earthly branches each in both the text of the manuscripts and the Nine Palaces diagram.[73] With the exception of Great Sound, the gods feature prominently in celestial journey verse and received ritual treatises.[74] Through their association with this set of deities, Harshness and Favor appear as deities themselves. Emissions of qi and the presence or absence of rain are the basis for sign-readings on days labeled Favor, Great Sound, and Harshness:

德發氣，大音雨之，嫁（稼）稅，雨吉。不雨，兵起，在軍，戰。
刑（刑）發氣，至德不雨，兵起，在軍，戰，雨吉。

> When on Favor, there is a release of qi, and on Great Sound it rains, in planting and harvesting, rain is auspicious. If it does not rain [on Great Sound], war will arise, and those in the army will go to battle.
>
> When on Harshness, there is a release of qi, and by Favor it does not rain, war will arise, and [those in] the army will go to battle. Rain is auspicious.[75]

Like other texts in the tomb, "Harshness and Favor" centers on the dual problems of agricultural production and military threats.

The meteoromantic portion of "Harshness and Favor" presents a series of sign-readings regarding the sun, moon, wind, rain, clouds, and qi. These reflect an abiding concern with warfare. Military woes dominate the discussion of halos about the sun and the moon. One passage suggests that ordinary soldiers perceived the dangers associated with certain types of halo phenomena. "When halos around the sun come into contact, the troops are afraid" 日交軍̠ (暈, 軍) 畏.[76] Observation of halos could potentially be used to argue against warfare, or to urge a ruler to seize military opportunities. Just as in the "Miscellaneous Sign-Readings," other signs involving halos could suggest global political change, such as the establishment of a new order of feudal lords. "When the sun is surrounded by nine-layered halos, the world will establish dukes and earls" 日軍 (暈) 九重, 天下有立公【伯】.[77] Military concerns and the criteria of color and height are also paramount in the observation of *qi*. *Qi* is relevant not only to military strategy but to broader political developments as well. One passage suggests that conquest can be predicted from the color of the sunset: "Look at the land. If the sun has already begun to set and one looks at it, and in its direction there is no cloud *qi*, yet it is a ruddy red color, the hearts of the people have wandered, and another (?) lord will possess them" 望 (望) 地, 日巳 (已) 入而望 (望) 之, 其鄉 (嚮) 無雲氣而康赤者, 民移心, 它 (?) 主有之.[78] In this case, the absence of *qi*, in the particular form of mist or fog, is a condition for the prognostic statement. *Qi* normalizes and nullifies the significance of a glow after sunset. In the immediately following passage, however, multiple appearances of odd *qi* phenomena at dusk portend devastation:

晦日望 (望) 氣若明而未明 (明), 兵取 (聚); 其鄉 (嚮) 若明 (明) 若日者, 其鄉 (嚮) 乃矍百軒。三見, 乃留 (流) 血苦 (枯) 骨。

If on the second-to-last day of the month one should observe *qi* that looks bright even before dawn, then soldiers will gather. If in a given direction it is as bright as the full sun,[79] then in that direction a hundred households will shudder in fear. Should it appear three times, then blood will flow and bones will bleach in the sun.[80]

The celestial signs of wind, rain, thunder and lightning, and *qi* in the "Harshness and Favor" manuscripts presaged events that would have deeply concerned regional courts in both the pre-imperial period and in the early Western Han.[81] The threat of war yet remained, and agricultural production would always be contingent upon the whims of the weather.

To appreciate the ways in which other manuscripts in the tomb might have shaped understanding of the celestial signs manuscripts, it will prove useful to first consider how the celestial signs manuscripts themselves conceptualized the relationship between Heaven and the human realm. To what extent did these manuscripts allow for human influence over the heavens? Were baleful signs in the heavens the result of human failures, such as ritual lapses or widespread suffering in the broader population? Did the celestial signs manuscripts take the view of Master Yan, who thought poor decisions led to baleful signs (and required changes in policy) or Duke Jing of Qi, who thought baleful signs appeared for no reason in particular, and required only a ritual response?

Causal direction is often difficult to determine, as the central concern of the manuscripts is identifying correspondences between signs and events. The manuscripts contain only the scantest of clues as to how actions in the human realm might cause baleful signs to appear, or might serve to expiate signs that have already appeared. Though the manuscripts relate human actions to the appearance of signs, they do not say whether the signs appear because of those actions, whether the signs influence human beings causing them to take those actions, or for that matter, whether some third factor might cause both signs in the heavens and actions in the human world. "Harshness and Favor" indicates a relationship between a nine-layered halo and the establishment of dukes and earls, but it does not say whether the nine-layered halo *causes* dukes and earls to be established, or whether the nine-layered halo *reflects* changes in the political order. The "Miscellaneous Sign-Readings" indicates a relationship between pairs of black rings and treacherous ministers, but it does not say if the sign *causes* ministers to become treacherous or if it merely *reflects* treachery in the human world. Developments in the human realm may cause the appearance of baleful signs, as Master Yan would have it, or such signs may appear for no particular reason to wreak havoc on the human realm, as Duke Jing seemed to believe.

There are some instances in which the appearance or meaning of a sign does seem to hinge on factors in the human realm. The presence or absence of rain on Great Sound days in "Harshness and Favor" respectively indicates either a bountiful harvest or military strife, though it is not difficult to imagine how rain might *cause* a better harvest, or how a poor harvest might lead to warfare. Jupiter's appearance in the celestial field corresponding to a particular domain in the "Five Planets" is more compelling. Depending on whether that polity possesses virtue *or not*, Jupiter's presence variably correlates with a bountiful harvest or stockpiling weapons. The planet might be interpreted as the cause of the agricultural or military developments, but could not have caused the presence or absence of virtue upon which those developments depended, since virtue is likewise treated as an independent variable. The correspondences between the Five Planets and celestial masters of aspects of court life such as music, rites, and sacrifice serve as a further clue as to how actions in the human realm might be tied to the appearance of celestial signs. These correspondences were *compatible with* arguments that failures to perform the correct rituals, music, or sacrifices at human courts led to the appearance of baleful signs in the heavens. Paradoxically, this might have supported a response to baleful signs using ritual expiatory means, such as those pursued by Duke Jing. Nonetheless, there is *no clear indication* in the manuscript as to how interpreters should conceptualize the relationship between the celestial masters of these affairs, signs involving the Five Planets, and corresponding actions in the human realm.

The "Five Planets," the "Miscellaneous Sign-Readings," and "Harshness and Favor" respond to and represent the human concerns of actors living in chaotic times. Threats of famine and flood, war and treachery, loom large in the texts. Charting the motions of celestial bodies is not an abstract goal but a response to disorder and uncertainty in the world. Yet the celestial signs manuscripts in themselves raise questions that remain unanswered. What role did sign-reading play in the broader enterprise of governance? Did baleful signs appear randomly or due to specific causes? What made for a successful or skillful reader of signs? How could disjunctures between prognostications and subsequent events be explained? The celestial signs manuscripts were one part of a larger corpus of texts that imposed an order on the cosmos and claimed that keeping with the cosmic order would in turn produce peace and prosperity in the human realm. Other texts in the tomb, including the *Constant Models* and manuscripts attached to the *Changes*, provide partial answers to these questions, emphasizing both the value of observing the heavens and the difficulty of interpreting signs.

Observation and Emulation in the *Constant Models*

The *Constant Models*,[82] like the *Daodejing*,[83] places rhetorical emphasis on adherence to cosmic models. These include Heaven (*tian* 天), the

Way (*dao* 道), and the Way of Heaven (*tian dao* 天道), terms that Jeffrey Richey has argued that the text uses interchangeably.[84] Law emerges from patterns inherent in these models. The sage holds fast to these patterns as he institutes a form of governance that mirrors the regularity of the cosmos.[85] The *Constant Models* argues that ideal rulers observe Heaven and emulate its movements. The ruler enacts the seasonal patterns of the heavens in the form of mutually opposed yet complementary modes of governance: harshness and favor, civil and military action (*wen* 文 and *wu* 武). Disasters emerge when rulers fail to emulate heaven, usually by acting with excessive violence. Baleful signs above and blatant misrule below justify and ensure the success of punitive campaigns against poor rulers. The text admonishes even just conquerors to practice restraint in the punishment of the vanquished.

In the *Constant Models*, Heaven embodies the virtues of constancy and reliability. To create order in his realm, the ideal ruler observes and emulates Heaven's model, instituting parallel virtues in his own domain. The sun is certain to rise and to set, the moon certain to wax and wane, and the stars certain to maintain their positions relative to one another as they endlessly revolve around the pole.[86] Heaven's model is articulated in the movements of celestial bodies and the changes of the seasons. The ideal ruler charts the cycles of the sun and moon, the stars and planets, and measures out the four seasons.[87] Careful measurement is paramount to following the model of Heaven and instituting it in the human realm:

輕重不稱, 是胃 (謂) 失道。天地有恆常, 萬民有恆事, 貴賤有恆立 (位), 畜臣有恆道, 使民有恆度。

If the light and the heavy weights are not balanced, this is called "Losing the Way." Heaven and Earth have enduring regularities. The myriad people have enduring affairs. Noble and base have enduring positions. There is an enduring way for caring for ministers, and there are enduring measures for employing the people.[88]

The ideal ruler mirrors the faithfulness, constancy, and enduring standards of Heaven as he rules the realm.

Emulating Heaven consists in enacting different modes of rulership at different times of year. The ideal ruler is a timely ruler, whose actions are modeled on the seasons themselves:

天有死生之時, 國有死生之正 (政)。因天之生也以養生, 胃 (謂) 之文, 因天之殺也以伐死, 胃 (謂) 之武。【文】武竝行, 則天下從矣。

Heaven has its seasons for meting out life and death. Domains have their policies for meting out life and death. Following Heaven's life-giving properties to nourish life; we call that *wen* (civil). Following Heaven's killing functions to attack and bring death, that we call *wu* (martial). When *wen* and *wu* are carried out in tandem, then the whole world surely obeys.[89]

Spring and summer are the time to bestow favors, while autumn and winter are the time to carry out punishments. The *Constant Models* prescribes an ethos of timeliness derived from the seasons of Heaven.

The *Constant Models* situates the kinds of disasters discussed in the "Five Planets," "Miscellaneous Sign-Readings," and "Harshness and Favor" as the products of specific failures, usually unjustified or excessive acts of violence. One passage explains: "In general, when a ruler violates the taboos and cuts himself off from the natural patterns in things, execution by Heaven will surely come" 凡犯禁絕理, 天誅必至.[90] Unjustified attacks on other states cause Heaven to send down calamities. A ruler who kills surrendered prisoners, punishes the innocent, or massacres submissive civilians inevitably meets with misfortune.[91] At the same time, the appearance of baleful celestial signs authorizes rulers to attack other states. Both *wen* and *wu*, civil and military means, belong to the seasons of Heaven and the repertoire of the ideal ruler. A just conqueror "follows the seasons of Heaven and punishes those whom Heaven would destroy" 因天時, 伐天毀, but relies on civil methods to a greater extent than military ones, reserving violence for the rare occasions that demand it. Should he depart from normative "human patterns" (*ren li* 人理), he himself is likely to come to a violent end.[92]

The *Constant Models* might be approximately described as a work of political thought, a how-to manual for the ruler who would align himself with cosmic principles. As such, it belongs to an entirely different genre than the celestial signs manuscripts. Nevertheless, the *Constant Models* and the celestial signs manuscripts belong to the same textual community, reflect a common worldview, and respond to a closely related set of problems. Developing a robust interpretation of the texts that addresses their archaeological context requires considering thematic relationships among them. Read together, the celestial signs manuscripts take on expanded significance.

An official who mastered both the *Constant Models* and the celestial signs manuscripts had the potential to provide an ethical basis for otherwise bare technical interpretations of baleful signs. Both human and cosmic factors could shape the outcome of a given course of events. Yates writes: "[W]arfare was treated as a rite, and was thought especially appropriate to divine about, because it involved matters of life and death in which the outcome, and very survival itself, was in the hands of the ancestors, gods, and spirits . . . in addition to being affected by human factors, such as the competence and morality of the general, the sufficiency of materiel, the level of training of the officers and the justice of the cause."[93] Whereas the celestial signs manuscripts aimed at the interpretation of cosmic factors, the *Constant Models* suggested a measure of human influence, consisting in both practical and ethical aspects. Disaster in the constellations corresponding to neighboring states might justify acts of violence against them, indicating not just that those states were in a position of weakness but that their rulers had transgressed against Heaven and failed to institute good governance. Disaster corresponding to one's own domain, on the other hand, could serve to authorize critiques of policies instituted by sitting rulers.

The *Changes* as a Guide for Reading Celestial Signs

Manuscripts related to the *Changes*, complete with numerous supplementary materials, were placed in the same lacquer box that contained the celestial signs manuscripts. These included a text closely corresponding to the "Tradition of the Appended Statements" (Xici zhuan 繫辭傳)[94] and several heretofore lost essays and dialogues, including the "Essentials" (Yao 要), the "Inquiries of Two or Three Disciples" (Er san zi wen 二三子問), and the eponymously titled "Muhe" 繆和 and "Zhaoli" 昭力.[95] The celestial signs manuscripts do not refer to these materials in any explicit way. However, both sets of materials were available to the same group of people at the same time. The Mawangdui *Changes* materials cast the celestial signs manuscripts in a different light. Scattered among the fragments, a picture emerges in which the *Changes* itself appears to be based on sagely emulation of celestial signs, even the best diviners often produce inaccurate results, and both auspicious and inauspicious portents demand immediate action.

The *Changes* materials cast the observation of celestial signs as a sagely enterprise. The opening of the text corresponding to the "Appended Statements" fittingly begins with an account of the differentiation of the cosmos. Mutually opposed yet complementary aspects of the cosmos, Heaven and Earth, high and low, movement and stillness, settle and produce stable patterns of difference: the pure *yang* hexagram Key and the pure *yin* hexagram River,[96] the noble and the base, the hard and the supple. In each of the cardinal directions, materials and living beings of the same kind gather. Birds form flocks, and beasts gather into herds. Good and ill fortune, images in the firmament, and the topography of the land emerge. The Eight Trigrams, the core of the *Changes*, produce celestial signs—thunder and lightning, wind and rain. The sun and moon both oppose and complete one another, emblems of constant, dynamic regularity.[97]

The sage himself makes the signs intelligible. "The sage set forth the hexagrams by observing the images, and he attached the line statements to them so as to illuminate good and ill-fortune" 耶〈聖〉人詆〈設〉卦觀馬〈象〉，殷〈繫〉辭焉而明〈明〉吉凶.[98] The hexagrams work as a convenient means of observing the signs that are already out there in the world, a simulacrum of images in the firmament. The *Changes* unlocks a way of knowing that is fully attuned to cosmic processes.[99] Based on sagely observation of both celestial patterns and earthly contours (*dili* 地理), the *Changes* is framed in the "Appended Statements" as a powerful tool for understanding developments in the human realm:

卬〈仰〉以觀於天文，頡〈俛〉以察於地理，是故知幽明〈明〉之故。觀始反〈返〉冬〈終〉，故知死生之說。

Gazing up they observed the celestial patterns, and lowering their heads they inspected the contours of the land, and thus they knew the reasons for obscurity and illumination. They observed beginnings and reflected on endings,[100] and thus knew how to explain life and death.[101]

The Mawangdui *Changes* materials, moreover, cast the sage ruler as one who is equally mindful of the state of the heavens and hearts of his subjects. One passage in the "Inquiries of Two or Three Disciples" reads:

聖人之立〈涖〉正〈政〉也，必尊天而敬眾，理順五行，天地無菑〈災〉，民口不傷，甘露時雨眾〈驟〉降，蘍〈飄〉風苦雨不至，民心相酳〈觴〉[102]以壽，故曰『番〈蕃〉庶』。

When a sage establishes his administration, he will for certain revere Heaven and be attentive to his subjects. He patterns himself on and follows the five kinds of action (*wu xing*), so that there are no disasters in Heaven or Earth. The people . . . are not harmed, so that sweet dew and timely rains fall in abundance, and whirlwinds and bitter rains do not arrive. The people toast to his long life in their hearts, thus [the *Changes*] says, "They prosper."[103]

The sage continually responds to both the heavens and his subjects, and both the heavens and his subjects respond to him. Auspicious signs appear above, promising good fortune, agricultural prosperity, and a well-fed, content populace.

The *Changes* materials likewise draw attention to the need to take timely action in response to signs, regardless of whether they appear in the firmament or as the result of milfoil casting. In another passage in the "Inquiries of Two or Three Disciples," Kongzi 孔子 (trad. 551–479), the sage figure better known in the West as Confucius, claims that the bottom *yin* (i.e., initial-six) line of the hexagram River, "treading on frost, the hard ice arrives" 履霜, 堅冰至, refers to "the timely admonitions of Heaven" 天時譜.[104] Through stalk casting, any given moment might call for a timely response—caution, trepidation, or steadfast endurance—that would otherwise be called for in the dead of winter. In the opening of "Muhe," an otherwise unknown disciple asks Kongzi the meaning of a line statement in the hexagram "Dispersion" (Huan 渙) that reads "Dispersion is to run up the stairs, regret vanishes" 渙賁 (奔) 其階, 每 (悔) 亡.[105] Kongzi explains that it is about seizing opportunities:

賁 (奔) 階, 幾也, 時也。古之君子, 時福至則進取, 時亡則以讓 . . . . . . 無千歲之國, 無百歲之家, 無十歲之能 . . . . . . 耵 (聖) 人知福之難得而賁 (奔) 也。

"Running up the stairs" refers to the moment of crisis, the opportunity. The lords of antiquity seized good fortune when it came along, and let it go when it vanished. . . . There are no domains that last a thousand years, no households that last a hundred, and no possibilities that last [even] ten. . . . The sage knows that good fortune is hard to come by and runs toward it.[106]

As Kongzi interprets it, the line statement points to the ephemeral nature of good fortune and the need to capitalize on it immediately.

The "Essentials" admits that milfoil casting carries a potential for spiritual corruption and often produces inaccurate results. In one passage, Kongzi's disciple Zigong 子貢 (d. 5th cent. BCE) asks him why he remains fond of the *Changes* despite the problems associated with milfoil casting.[107] Kongzi acknowledges the dangers of casting stalks without care or good character, but identifies the *Changes* as a text that contains ancient lessons traceable to King Wen 文 (d. 1046), the spiritual founder of the Zhou dynasty. Where other texts, such as the *Documents*, are incomplete, the *Changes* is more or less intact. Kongzi explains:

察亓 (其) 要者, 不趮 (詭∖悗) 亓 (其) 辤 (辭)。《尚書》多於 (闕),《周易》未失也, 且又 (有) 古之遺言焉。予非安亓 (其) 用也。

Those who clearly understand its essentials do not alter its words (*ci* 辭). There is much that is missing (*yu* 於) from the *Venerable Documents*, but no part of the *Changes of Zhou* has been lost.[108] Moreover, it contains the remnant words of antiquity. It is not the case that I am at ease in using it.[109]

Kongzi tells Zigong that the *Changes* itself is a work invested with the virtue of King Wen,[110] whom legend credited with constructing the Sixty-four Hexagrams of the *Changes* from the existing Eight Trigrams.[111] Discussing the hexagrams Gain (Yi 益) and Loss (Sun 損), Kongzi holds up the *Changes* as a means of understanding the cosmos that goes beyond the stars: "The *Changes* has the Way of Heaven in it, and even the sun, moon, planets and stars do not fully measure up to it" 故《易》又 (有) 天道焉, 而不可以日月、生 (星) 辰盡稱也.[112] Kongzi likewise claims that the *Changes* contains the Way of Earth (*di dao* 地道), which amounts to more than water, fire, metal, soil, and wood, and the Way of the Human Realm (*ren dao* 人道), which amounts to more than the relationships among family members or lords and ministers. A complete system for understanding the whole of the cosmos, the *Changes* exceeds even rites, music, the *Odes*, and the *Documents* in utility for good governance. The *Changes* contains the Way of the Lord (*jun dao* 君道).[113]

The Mawangdui *Changes* materials call for caution and restraint in the use of milfoil casting. The "Essentials" explains that results from milfoil casting are unreliable, even when a person of good character carries out the procedure. The text of the *Changes* embodies principles both cosmic and human. Acting on the basis of these principles is the path to ensuring prosperity and stability. Casting stalks is a fraught process and rarely necessary:

君子德行焉求福，故祭祀而寡也；仁義焉求吉，故卜筮而希（稀）也。

A gentleman seeks good fortune by acting with virtue, and so, he seldom conducts sacrificial rites. He seeks good auspices through humaneness and duty, and so, he rarely casts the stalks.[114]

Sign-reading is to be performed in rare moments of uncertainty, as a final recourse when ordinary means of decision-making prove inadequate or inconclusive, but its results are themselves highly uncertain. Kongzi explains to Zigong: "Out of one hundred sign-readings, I get it right [only] seventy times" 子曰：吾百占而七十當.[115] The character of the person using the *Changes* is all-important: "Without virtue (*de*), one cannot understand the *Changes*, and so, the gentleman is reverent towards it" 無（?）德，則不能知易，故君子奠（尊）【之】.[116] Even the most virtuous and most skilled practitioners of sign-reading could not hope to approach perfect accuracy.

Like the *Constant Models*, the *Changes* is different in kind from the celestial signs manuscripts, though it too explicitly concerns the reading of signs. Attached materials, including both the "Appended Statements" and heretofore lost documents interred in the tomb, illuminate the otherwise obscure process of milfoil casting. By extension, these materials may apply to processes of sign-reading in general. The "Appended Statements" provides a cosmological basis for milfoil casting, notably rooted in the sages' observations of *tianwen*, patterns in the heavens. Texts such as the "Inquiries of Two or Three Disciples" and "Muhe" emphasize the import of timely response to observed signs and conditions. The "Essentials" suggests milfoil casting should rarely be performed, is fraught with uncertainty, and should be carried out only by the virtuous. The presence of these texts in the same community as the celestial signs manuscripts established the possibility for making similar claims about other types of sign-reading.

Virtue, Heaven, and the *Five Kinds of Action*

The *Changes*' emphasis on *de*, or virtue, raises a further question: What did *de* mean in the context of the Mawangdui corpus? Of all the texts entombed at Mawangdui, the *Five Kinds of Action* and its commentary offer the most pointed discussion of *de*.[117] The text's opening section describes *de* as a master virtue that emerges among those who engage in all five kinds of (virtuous) action:

【.仁】刑（形）【於內】胃（謂）之德之行 . . . . . .
知（智）刑（形）於內胃（謂）之德之行 . . . . . .
【義刑（形）】於內【胃（謂）】之德之行 . . . . . .
禮刑（形）於內胃（謂）之德之行 . . . . . .
聖刑（形）於內【胃（謂）之德之行 . . . . . .
德之行五，和胃（謂）之德;
四行，和胃（謂）之善＿（善。善），人道也; 德，天道也。

When humaneness forms within, this is called virtuous action . . .
When wisdom forms within, this is called virtuous action . . .
When duty forms within, this is called virtuous action . . .
When the ritual propriety forms within, this is called virtuous action . . .
When sagacity forms within, this is called virtuous action . . .
When five kinds of virtuous action harmonize, this is called *de* (virtue).
When four kinds harmonize, this is called goodness.
Goodness is the Way of the Human Realm, *de* is the Way of Heaven.[118]

The standard for true *de* is high. It requires a ruler to develop and act out of five qualities: humaneness, wisdom, duty, ritual propriety, and sagacity. If a single one of these qualities is absent, the ruler may be good (*shan* 善), but does not possess *de*. Possessing true virtue means the ruler has exceeded the normative standards for humanity, fully realizing the Way of Heaven.

The *Five Kinds of Action* scarcely mentions sign-reading. Most knowledge comes through seeing, by hearing examples, or by way of analogy.[119] In a single instance, however, the text describes understanding the trajectory of events as a special form of perceptiveness that belongs to Heaven, or by extension, to those of such virtue that they embody the Way of Heaven: "To know it by examining the auspices/the smallest of things (*ji*) is Heavenly" 鐖 (幾) 而知之, 天也. Pang Pu 龐樸 interprets this line as a reference to the technical process of reading signs, glossing *ji* as *ji* 機 (examining the auspices).[120] Other scholars read *ji* in the same sense as the parallel character in the recension of the text from the Guodian 郭店 archaeological site (ca. 300 BCE), *ji* 幾 (the smallest of things). In either case, extraordinary powers of perception belong to those whose *de* matches Heaven.

The *Five Kinds of Action* and its commentary position the ideal ruler as one who follows the model of Heaven and extends that model so that it can be understood and applied. He follows Heaven's patterns, and according to the commentary, acts as King Wen once did, creating a pattern for his subjects to follow.[121] The final line of the main text suggests that such rulers exhibit a spontaneous, emotional response to the Way: "Those who have virtue (*de*) feel joy when they hear of the way" 聞道而樂, 有悳 (德) 者也. The commentary to the text ties the Way to the movements of the heavens: "The Way is the Way of Heaven" 道也者, 天道也.[122] *Five Kinds of Action* is a text about embodying, exhibiting, and acting with *de*. *De* is not one virtue among many but the cardinal virtue that manifests only when all others are in place. As a general rule, *de* need not point to ethical or moral qualities—it may simply indicate power or charisma. Yet *Five Kinds of Action* does not permit such a reading. A ruler cannot possess *de* who does not act with humaneness, among other admirable qualities. If readers interpreted their texts as a corpus, this sense of *de* likely influenced their understanding of *de* in the *Changes*, the *Constant Models*, and even the celestial signs manuscripts.

Conclusion

The texts in the Mawangdui corpus doubtless entered into the worldview of those who read and used them. Yet their fragmentary nature allows for only limited conclusions regarding their usage and interpretation. Treating the texts as a corpus does not show how they were actually interpreted or what connections their users in fact drew among them. Reading the manuscripts together does, however, bring to light common thematic elements and concerns that ran through seemingly disparate materials.

The celestial signs manuscripts tied specific types of signs to problems of food security, treachery at court, military conflict, and instability in the broader political order. They established correspondences that paired clouds of certain shapes and specific constellations with particular states. The celestial signs manuscripts allowed their users to determine *what* signs indicated but provided little guidance concerning the causal question of *why* baleful signs came into being or the pragmatic question of *how* they should be addressed. The old debate between Master Yan and Duke Jing remained. Reading the celestial signs manuscripts in isolation, one could take Duke Jing's position that baleful signs came into being for mysterious reasons and could be expiated by ritual means, though the manuscripts did not explicitly contradict Master Yan's position that baleful signs arose due to failures in governance and could be addressed only by rectifying those failures. In what ways might the *Constant Models*, *Changes*-related materials, and the *Five Kinds of Action* have proved germane to this debate, and more broadly, provided the textual community at Mawangdui with sources of authority for making claims about the relationship between Heaven and the human realm?

The *Constant Models* injects the notion of kingship into the observation of the movements of celestial bodies. It emphasizes the critical importance of emulating their patterned movements, their constancy and regularity, and enacting these through good governance. Creating a kind of celestial order in the human world, the ideal ruler measures the movements of Heaven, following its seasons as he variably employs civil and martial methods, carefully meting out rewards and punishments, life and death. The text explicitly treats baleful signs as a result of excessive or unnecessary violence, poor governance that departs from the Way of Heaven. It allowed for interpreting such signs as Master Yan did, for claiming they resulted from specific failures that demanded correction.

The *Changes* materials contain a range of ideas that might likewise be applied to questions of why signs appeared and how they should be addressed. Their broader significance extended well beyond milfoil casting. The "Appended Statements" provides a detailed, cosmological rationale for identifying correlations between *yinyang* dynamics in the heavens and in the human world. More generally, it allows for framing the appearance of baleful signs in terms of resonances between human beings and the cosmos. It claims that sage rulers observe and reflect on cosmic patterns including both *tianwen* and *dili* (the contours of the land), thereby acquiring their mysterious power to decipher patterns of change. "Inquiries of Two or Three Disciples" and "Muhe" emphasize the critical import of taking stock of current circumstances and the need not only to gather knowledge but also to act upon it without delay. Finally, while the celestial signs manuscripts in themselves do not provide a way of explaining why mantic practices so often proved unreliable, the *Changes*-related "Essentials" manuscript works to normalize error. Sign-readers in the ancient world doubtless encountered numerous occasions when their art seemed to fail. The "Essentials" told them, on the one hand, that such occasions were to be expected, and on the other, that texts such as the "Five Planets" and "Miscellaneous Sign-Readings" would be of no use to them if they lacked *de*.

Five Kinds of Action provides a working definition of *de* as a virtue of the highest order, one that encompasses and subsumes humaneness, wisdom, duty, ritual propriety, and sagacity. For those who took to heart both this definition of *de* and the "Essentials" claim that *de* was necessary for performing successful sign-reading, the celestial signs manuscripts were not mere technical texts, usable by anyone who could read them, but ritual objects that gained power only when used by the right people for the right purposes.

Despite the disparate nature of the texts in the tomb, their presence within a common textual community made it possible to draw connections between them. Omenology and imperatives for good governance existed side by side, even if they did not interpenetrate on a textual level. Later in the century, as the empire became more stable and centralized, other courts collected small units of sundry texts, integrating technical knowledge into large-scale projects that combined that knowledge with statecraft, cosmological theories, political thought, and narratives concerning exemplary figures, both historical and legendary. The following chapter turns to two of the greatest of these works, the *Huainanzi* and the *Shiji*.

Chapter 2

Sign-Reading and Sagely Rulership in the *Huainanzi* and *Shiji*

Liu An's *Huainanzi* (submitted to the throne 139 BCE) and Sima Qian's *Shiji* (compiled ca. 100 BCE) are the earliest major textual projects that incorporate technical omenological information into a broader discourse about good governance. Each contains a treatise on the heavens consisting primarily of information similar in kind to that found in excavated technical manuscripts. The *Huainanzi* and *Shiji*, however, also make explicit statements regarding the value of that information that extend beyond their technical treatises. The *Huainanzi* includes cosmological theories and scattered narratives that demonstrate the value of skillful sign-reading, while the *Shiji* "Basic Annals" (*ben ji* 本紀) position the ability to read signs well as a characteristic that distinguishes the best rulers from the worst. Following a brief overview of the historical circumstances under which the *Huainanzi* and *Shiji* were compiled and the technical features of their treatises on celestial signs, this chapter turns to the cosmological theories and narrative accounts that demonstrated the value of sign-reading and related technical knowledge for sagely governance.

Historical Circumstances and Technical Content

Both the *Huainanzi* and the *Shiji* were compiled during the reign of Emperor Wu. Though the *Huainanzi* was submitted to the throne less than thirty years after the Mawangdui tombs had been sealed, the empire had become stable and centralized. Emperor Jing 景 (r. 157–141) had crushed the Seven Kingdoms Rebellion of 154, consolidating imperial power. Liu An's father, a son of the founding emperor, had died en route to exile after fomenting rebellion in 174. Liu An inherited a restored but reduced Kingdom of Huainan in 164. In 141, the fifteen-year-old Emperor Wu took the throne. Two years later, Liu presented him with a guide for statecraft that encouraged adherence to the cosmic order and minimal intervention in regional affairs.[1]

Emperor Wu, however, proved determined to expand his empire and to further strengthen the power of the central court over regional rulers. Over the course of the next two decades, he engaged in offensives against the Xiongnu in the north, attempted to establish a commandery in the Korean peninsula, and sent emissaries as far west as Bactria.[2] The fight against the Xiongnu played a critical role in the lives of both Liu An and Sima Qian. After Liu An's son Liu Qian 劉遷 (d. ca. 122 BCE) was brought up on charges for preventing his former client Lei Pi 雷被 (fl. 124) from enlisting to fight the Xiongnu, Liu An's kingdom was reduced in size. The following year, figures at court who advocated concentrating power in the hands of the emperor, led by Superintendent of Trials (*ting wei* 廷尉) Zhang Tang 張湯 (d. 116 BCE) and Chancellor Gongsun Hong 公孫弘 (ca. 200–121), investigated Liu An for plotting a rebellion. In 122, he was found guilty and executed.[3] In Sima Qian's case, the desire to complete his textual project kept him going after he suffered a humiliating punishment ordered by Emperor Wu. Following his father's death in 110, Sima continued the project until sometime after 99, when he ran afoul of Emperor Wu for defending the general Li Ling's 李陵 (d. 74) decision to surrender to the Xiongnu. Faced with the prospect of either execution or castration, Sima chose life so that he could complete his father's work, thereby carrying on his lineage through text if not through flesh.[4]

The two treatises contain overlapping technical materials, but they differ substantially in both organization and content (see boxes 2.1 and 2.2). The *Huainanzi* treatise is more heterogeneous and broader in scope than is the *Shiji* "Celestial Offices" or, for that matter, the *Hanshu* "Celestial Patterns." It is highly inclusive, presenting an array of technical information likely derived from multiple sources.[5] It begins with an overview of cosmology, describing the mechanisms that produce living creatures and the natural world, and that underlie the resonances between human beings and celestial phenomena. Its next section presents

Box 2.1
Organization of the *Huainanzi* "Tianwen" Treatise[6]

Cosmology

Formation of the world
Yin and yang
Universal resonance and human government

Stars and Celestial Divisions

The Nine Fields 九野
The Five Planets 五星
The Eight Winds 八風
The Five Offices and Six Bureaus 五官六府
The Court of Taiyi 太一之庭

The Sun, the Moon, and the Calendar

The Northern Dipper 北斗 and the solar year
Quarter-remainder Calendrical System 四分法
The Seven Stations of Favor and Harshness 刑德七舍
The Twenty-four Solar Periods 二十四節氣
The Little Year 小歲, the Grand Year 大歲 and the Celestial Cords 天維
Luni-solar intercalation
The calendrical cycle of the *wuxing* 五行
The gods and the cycle of seasons
The daily path of the sun and divisions of the day

Calendrical Harmony and Meteorology

The Northern Dipper and the Twelve Pitchpipes 十二律
The generation of pitchpipes and their harmonic numbers
Meteorology and calendrical harmony

The Grand Yin, Calendrical and Hemerological Usages

The triple revolution of the Grand Yin 太陰
The method of Establishing and Ousting 建除
The Grand Yin and the rule of Assistant Conductors 攝提格
The method of Favor and Harshness and the Grand Yin
The Twenty-Eight Starry Lodges 二十八星宿
Prognostics of the Year Star 歲星 (Jupiter) and the Grand Yin
The Ten Stems and the Twelve Branches 十干十二支
The method of Canopy and Chariot 堪輿
The unity of the natural and human worlds
The attendants to the year and the rule of Assistant Conductors

The Gnomon and the Dimensions of the World

The east–west axis and the tropical movement of the sun
The site of observation and the measurement of terrestrial distances
Calculating the height of the sky over the earth

Box 2.2
Organization of the *Shiji* "Celestial Offices"

The Stars

Central Palace 中宮 (Circumpolar Region)
Eastern Palace 東宮
Southern Palace 南宮
Western Palace 西宮
Northern Palace 北宮

The Five Planets

Jupiter (Year Star 歲星)
Mars (Dazzling Deluder 熒惑)

Saturn (Quelling Star 填星)
Planetary interactions
Venus (Great White 太白)
Mercury (Watch Star 辰星)

Regional Field Allocations

The Sun and the Moon

Emergent "Stars"

Clouds and *Qi*

Tetrasyllabic Poem

Divining the Harvest

Sima Qian's Postface

Basis in cosmology and early sign-reading
Decline of Western Zhou
Chunqiu and Warring States eras
Twenty-eight lodges and twelve regions
Qin conquest and collapse
Founding of Han and early Han experts
Proper responses to disasters

materials resembling those found in later *tianwen* texts. It names and categorizes asterisms (albeit in considerably less detail than do later treatises), describes the motions of the planets, and identifies their correspondences with the human world and broader cosmos. Much of the main body of the treatise, however, focuses on materials that are similar in kind to those found in the *Shiji* and *Hanshu* treatises on calendrics and harmonics (*lüli* 律曆/歷).[7] These include complex descriptions of the position of the Northern Dipper at various times during the solar year, methods for calculating the relationship between the lunar and solar year, gods associated with seasonal cycles, and the relationship between the Northern Dipper and the Twelve Pitchpipes (and their musical modes), among numerous other items. The treatise likewise includes hemerological knowledge, establishing correspondences between particular hours, days, times of year, and cosmic conditions. The *Huainanzi* treatise is sometimes inconsistent, citing, for instance, three different lengths for the year. Parts of the treatise bear comparison to the Mawangdui "Harshness and Favor" manuscripts, including their meteoromantic sections, and to the cycle of Jupiter in the "Five Planets Sign-Readings."[8] While the calendrical and hemerological aspects of the treatise largely fall outside the focal range of the present study, it should be noted that the *Huainanzi* "Tianwen" treatise itself does not assign these to separate categories. The treatise as a whole represents a broad, early conception of *tianwen* that contrasts with the somewhat narrower usage of the term that took hold by the late Western Han.

The *Shiji* treatise, for its part, opens by describing the names and positions of stars and constellations, including the circumpolar stars most closely associated with the imperial court and each of the twenty-eight lunar lodges. As David Pankenier has shown, it not only details which parts of the sky map onto which parts of the empire but also divides the sky into *yang* and *yin* portions, respectively corresponding to the empire itself and to distinct cultural groups beyond its northern and western frontiers.[9] It provides detailed descriptions of the movements of the Five Planets and their omenological significance, recording technical information on transient and meteorological phenomena including strange clouds, rainbows, halos around the sun and moon, and emergent "stars" such as comets, novae, and meteors. It also includes a tetrasyllabic poem on disastrous weather and a section on divining the harvest at the beginning of the new year. Sima Qian concludes with an essay that records historical instances of celestial signs and explains the value of the technical materials collected in the treatise.[10]

Examining the correlations the two treatises make between the Five Planets and other aspects of the cosmos shows the character of their technical, omenological differences (see tables 2.1 and 2.2).[11] The *Huainanzi* correspondences are more similar to those in the "Five Planets" manuscript than they are to those in the *Shiji*. Like the Mawangdui text, the *Huainanzi* treatise lists a set of correspondences among gods, assistants, and the Five Planets. These are absent in the *Shiji*. The *Huainanzi* treatise also includes correspondences with five tools and five tones found in

Table 2.1. Planetary Correspondences in the Huainanzi "Tianwen" Treatise

	Jupiter	Mars	Saturn	Venus	Mercury
Direction	East	South	Center	West	North
Material	Wood	Fire	Soil	Metal	Water
God	Tai Hao 太皞	Fiery God 炎帝	Yellow God 黃帝	Shao Hao 少昊	Zhuan Xu 顓頊
Head of the Left 左	Goumang 句芒	Zhu Ming 朱明	Lord Soil 后土	Ru Shou 蓐收	Xuan Ming 玄冥
Tool	Compass 規	Balance Beam 衡	Plumb Line 繩	Square 矩	Weight 權
Season	Spring	Summer	Four Directions 四方	Autumn	Winter
Spirit	Year Star 歲星	The Dazzling Deluder 熒惑	Quelling Star 鎮星	Great White 太白	The Watch Star 辰星
Animal 獸	Azure Dragon 蒼龍	Vermilion Bird 朱鳥	Yellow Dragon 黃龍	White Tiger 白虎	Dark Warrior 玄武
Tone 音	jiao 角	zhi 徵	gong 宮	shang 商	yu 羽
Days	jia, yi 1–2	bing, ding 3–4	wu, ji 5–6	geng, xin 7–8	ren, gui 9–10

neither the excavated text nor the *Shiji* treatise. The *Huainanzi* treatise alone includes correspondences with the four animal constellations that encompass the twenty-eight lunar lodges, as well as a "Yellow Dragon" corresponding to Saturn, perhaps identifiable as the circumpolar stars. The *Shiji* treatise exhibits more restraint in what it includes. It uses the same directions, materials, seasons, and days as the *Huainanzi* and Mawangdui "Five Planets" manuscript. It jettisons all other categories of correspondence but adds one of its own that ties planetary anomalies to ethical failures. Errors related to duty and ritual propriety correspond to spring and summer, Jupiter and Mars. Errors related to reduction[12] and

Table 2.2. Planetary Correspondences in the *Shiji* "Celestial Offices"

	Jupiter	Mars	Saturn	Venus	Mercury
Location Method	Examine the paths of the sun and moon to measure forward or retrograde motion of the Year Star. 察日、月之行以揆歲星順逆。	Examine rigid *qi* to locate the Dazzling Deluder. 察剛氣以處熒惑。	Calculate its conjunction with Dipper to determine the position of the Quelling Star. 曆斗之會以定填星之位。	Examine the path of the sun to locate the position of Great White. 察日行以處位太白。	Examine the conjunction between the sun and the Watch to set to order the position of the Watch Star. 察日辰之會，以治辰星之位。
Direction	East	South	Center	West	North
Material	Wood	Fire	Soil	Metal	Water
Season	Spring	Summer	Late Summer 季夏	Autumn	Winter
Days	*jia, yi* 1–2	*bing, ding* 3–4	*wu, ji* 5–6	*geng, xin* 7–8	*ren, gui* 9–10
Punishment for the loss of . . .	Duty 義	Ritual Propriety 禮	When ritual, favor, duty, reduction, and harshness all are lost, only then does the Quelling Star waver. 禮、德、義、殺、刑盡失，而填星乃為之動搖。	Reduction 殺	Harshness 刑

harshness correspond to fall and winter, Venus and Mercury. Saturn, a paragon of stability with its near thirty-year journey through the stars, wavers only if these four and favor (*de*) are all lost.

Tianwen and the *Huainanzi* "Tianwen" Treatise

The *Huainanzi* "Tianwen" treatise occurs within a broad syncretic text produced at the court of the King of Huainan that promoted, as Major et al. write, "a Zhou style realm in which royal relatives administered semi-autonomous local kingdoms while giving their ultimate allegiance to the benevolent rule of an enlightened sage-emperor."[15] The *Huainanzi* contains a survey of knowledge needed for successful governance. Using a "root and branch" organization, it lays out the nature of the cosmos and of human societies in its early chapters and discusses more practical matters in its later ones. The complete work provides "a model curriculum for a monarch-in-training."[16] The *Huainanzi* effectively weaves the type of technical discourse seen in the Mawangdui celestial signs manuscripts and the ethical discourse on good governance in the *Constant Models* manuscript into a single textual project.[17]

The exhaustive technical information in the *Huainanzi* "Tianwen" treatise enhances the overall rhetorical effect of the work as a whole, though the political ideals the larger text advocates, including its polemics[18] in favor of a decentralized, restrained mode of governance, remain in the background of the treatise. The *Huainanzi* "Summary of the Essentials" (Yao lüe 要略) explains, as Major et al. note,[19] the role the treatise plays in the overall project of governing an empire:

《天文》者, 所以和陰陽之氣, 理日月之光, 節開塞之時, 列星辰之行, 知逆順之變, 避忌諱之殃, 順時運之應, 法五神之常, 使人有以仰天承順, 而不亂其常者也。

The (Teachings on) Celestial Patterns are the means by which His Majesty[20] may be in harmony with the *qi* of *yin* and *yang*, pattern[21] the light of the sun and moon, follow the rhythms of the seasons of opening and obstruction, sequence the movements of the planets and constellations, understand the transformations in their retrograde and proper motions, avoid the dangers of violating the taboos, follow the appropriate responses to the advance of the seasons, emulate the constant principles of the Five Spirits (i.e., the Five Planets), and make it so that people will have a way to gaze up at Heaven and be able to follow it, so as not to violate its constant principles.[22]

Observation of celestial signs and emulation of celestial patterns play a central role in good governance.

Certain passages in the "Tianwen" treatise itself also draw attention to the interaction between the human realm and the cosmos. Ineffectual, ritually inappropriate, or unjust governance produces celestial signs, some of which may have severe consequences for agricultural production: "If there are punishments and violence, there will be whirlwinds. If there are wrongful laws and ordinances, there will be plagues of devouring insects. If there are unjust executions, the land will redden with drought. If there are unseasonable ordinances, there will be a great excess of rain"[23] 故誅暴則多飄風, 枉法令則多蟲螟, 殺不辜則國赤地, 令不收則多淫雨. Drought and floods lead to a hungry and rebellious populace. Such signs have direct implications for the ability of the ruler to maintain power.

Qi and the Integrated Cosmos

Cosmography in the *Huainanzi* creates a palpable sense of how different parts of the cosmos interact, including the human and celestial realms. Elaborate descriptions of *qi* in the "Tianwen," "Jingshen" 精神 (Essential Spirit), and "Taizu" 泰族 (Exalted Lineage) treatises delineate a cosmos in which the body, the earth, and the stars influence and resonate with one another. This vision of the cosmos establishes a powerful theoretical basis for claims that the heavens respond to the actions of human rulers.

The "Tianwen" treatise opens by describing the separation of primordial *qi* (*yuan qi* 元氣) into "clear yang"[24] 清陽 *qi* that forms the heavens and "heavy and turbid" 重濁 *qi* that forms the earth. These two types of *qi* engage in a dynamic process that gives rise to the four seasons, the sun, the moon, planets, and stars.[25] Living beings also emerge from *qi*. Animals with feathers and fur that fly or walk with their bodies separate from the earth are essentially *yang*, while those with scales or shells that burrow into or slither along the earth are essentially *yin*.[26] The sun and the moon influence the life cycles of creatures through sympathetic *yang* or *yin* relationships:

日者, 陽之主也, 是故春夏則群獸除, 日至而麋鹿解。月者陰之宗也, 是以月虛而魚腦減, 月死而蠃蛖膲。

The sun is the ruler of *yang*. Therefore, in the spring and summer animals shed their fur; at the summer solstice, stags' antlers drop off. The moon is the fundament of *yin*. Therefore when the moon wanes, the brains of fish shrink; when the moon dies, wasps and crabs shrivel up.[27]

The treatise integrates winds and floods, clouds and comets, meteors and eclipses into a network of relationships. Categorical correspondences between objects of like kind explain movement, change, and aberrant phenomena. Undergirded by *qi*, this network links together living creatures with the whole of the cosmos. Animals do not merely play a passive role within it, but their activities produce emanations of *qi* responsible for the appearance of celestial signs. Dragons produce clouds, *qilins* eclipses, and leviathans comets.[28]

The "Tianwen" treatise explains inclement weather as a function of *qi*. While the treatise does not associate *qi* with emissions from a pantheon of gods, *qi* that is properly balanced does indeed produce seasonable weather, whereas imbalances in *qi* result in inclement weather—weather that is bad for business, in one way or another: "When *yin qi* is dominant, there are floods; when *yang qi* is dominant, there is drought" 陰氣勝則為水, 陽氣勝則為旱.[29] Changes in *qi* produce weather of all types. Deviations from regular cycles cause storms. If *yang qi* prevails, rain falls, and if *yin qi* prevails, there is snow and frost.[30] Periods of both sun and rain are necessary for agriculture to be successful. Frost is harmful if it occurs in the summer, both because it is a sign that something is awry with the cosmos, and more immediately, because it destroys crops. Timing is crucial.

The ruler must act in a manner befitting the season. The "Tianwen" treatise prescribes actions to be undertaken during five seasons, each seventy-two days long and tied to the sexagenary calendar. During the first season, which begins on a *jiazi* (1) day, the ruler should "act gently and graciously, relax the many prohibitions, open doors and covers, remove barriers, and avoid cutting down trees" 行柔惠, 挺群禁, 開閉扇, 通障塞, 毋伐木.³¹ The appropriate actions of government are a function of the cosmic conditions that prevail at a given time. Carrying out actions at the wrong time results in cosmic imbalances, leading insect larva to hatch too early, women to suffer miscarriages, spring frosts to fall, and silk worms to fail to mature. There may be earthquakes, cold spells in summer, poor harvests, and droughts.³² In a world where the ruler stands at the center of the cosmos, there are consequences for untimely actions. The ruler and his court are responsible for cosmic disturbances.

Heaven and the Human Heart

The *Huainanzi* "Tianwen" treatise claims that "the dispositions of the rulers of the human realm penetrate the heavens" 人主之情, 上通于天.³³ The relationship between microcosm and macrocosm, the ruler and the world, manifests through the dynamic actions of *qi*. The "Jingshen" treatise proposes a model in which human bodies and the cosmos itself are in a state of constant flux, so that celestial signs emanate from the viscera of the ruler and his subjects. Its opening passage claims that human beings receive their spirits from Heaven and their bodies from Earth, positioning them as active participants in the cosmos whose actions and emotions produce celestial signs:

> 天有風雨寒暑, 人亦有取與喜怒。故膽為雲, 肺為氣, 肝為風, 腎為雨, 脾為雷, 以與天地相參也, 而心為之主。

> Heaven has wind, rain, cold, and heat; humans have taking, giving, joy, and anger. The gall manifests (*wei*) clouds, the lungs manifest *qi*, the liver manifests wind, the kidneys manifest rain, and the spleen manifests lightning, and so, they form a triad with Heaven and Earth, and their hearts serve as its ruler.³⁴

The human heart, as center of both emotion and cognition, occupies the position of the ruler and is the ultimate source from which celestial signs issue. The body becomes a cosmos, and the cosmos a body, so that irregularities in the body explain disaster and calamities:

> 是故耳目者、日月也, 血氣者、風雨也 . . . . . . 日月失其行, 薄蝕無光; 風雨非其時, 毀折生災; 五星失其行, 州國受殃。

Therefore, the ears and eyes are the sun and moon; the blood and *qi* are the wind and rain. . . . When the sun and moon err in their periodic motions, fireflies have no light, wind and rain come out of season, destruction occurs and disasters arise. When the Five Planets do not follow their courses, provinces and domains meet with calamity.[35]

In the "Jingshen" treatise, visceral excretions of the ruler, members of his court, and his multitudes of subjects all produce baleful signs. Other passages in the *Huainanzi*, however, emphasize the ruler's personal responsibility to emulate the ideal order of Heaven and to be watchful for celestial signs.

The "Taizu" treatise positions the sage as a constant and perspicacious observer of the cosmos, one who "embraces the heart of Heaven" 懷天心.[36] The sage remains hidden from view, modeling his actions on the regularities of the cosmos,[37] interfering little in the everyday business of the empire. He reigns rather than rules, possessed of a transformative power that produces auspicious signs and excellent agricultural conditions:

故聖人者懷天心，聲然能動化天下者也。故精誠感於內，形氣動於天，則景星見，黃龍下，祥鳳至，醴泉出，嘉穀生，河不滿溢，海不溶波。故《詩》云：「懷柔百神，及河嶠岳」。

When the Sage embraces the heart of Heaven, the sound of his voice is capable of transforming the world. Thus, when his quintessential integrity stirs within, forms of *qi* move in Heaven, and so lucky stars appear, yellow dragons descend, auspicious winds[38] arrive, sweet springs emerge, fine grains thrive, rivers do not fill and overflow, and the seas do not churn and roil. Thus the *Odes* says, "Nurturing and yielding are the hundred spirits, even to the rivers and mountain peaks."[39]

The sage transforms the cosmos not by imposing particular policies, laws, or even rituals, but by patterning his own heart on the heart of Heaven. So long as the ruler's own heart mirrors the cosmic heart, order prevails in the world. Should he instead oppose Heaven and shirk his responsibilities, eclipses occur, the Five Planets stray from their courses, rivers flood, mountains crumble, and summer frosts fall.[40] The text points to a fundamental inseparability of the heavens and the human realm:

"Heaven and humanity mutually flow into one another. Thus when a domain is endangered and perishes, celestial patterns aberrate" 天之與人有以相通也。故國危亡而天文變。[41] Just as baleful signs in the heavens precede the destruction of a polity, human actions precede baleful signs.

The *Huainanzi* supports these general claims by recalling historical narratives in which exemplary rulers read and respond to baleful signs. One of the most memorable emphasizes the virtue of self-sacrifice, while another shows that signs of the fall of one dynasty also signal the rise of another.

The first story concerns an appearance of the Dazzling Deluder, Mars, in the constellation corresponding to the domain of Duke Jing of Song 宋景公 (r. 516–451). Zi Wei 子韋, an expert in reading celestial signs, explained that the Duke could transfer the baleful influence of the planet to his prime minister, his people, or the harvest. Duke Jing refused, choosing to sacrifice himself instead. Mars moved three lodges in response to his virtue, and he lived another twenty-one years.[42] The story became a staple of Han rhetoric, particularly memorials in which writers called on rulers to show leniency to their subjects or high officials.[43]

The second narrative presents King Wu 武 (r. 1046–1043 BCE) as a wise and virtuous reader of baleful celestial signs:

武王伐紂，東面而迎歲，至汜而水，至共頭而墜，彗星出而授殷人其柄。當戰之時，十日亂於上，風雨擊於中，然而前無蹈難之賞，而後無遁北之刑，白刃不畢拔而天下得矣。

When King Wu attacked Zhow, he faced east and welcomed the year. When he reached the Si River, there was a flood; when he reached Gongtou [Mountain], it collapsed. A broom-star appeared and presented its handle to the men of Yin. During the battle, ten suns rioted above and wind and rain struck below. Yet, in front there were no rewards for braving danger, and at the rear, no punishments for flight. Clean blades were never fully drawn and the world submitted to him.[44]

These signs occurred at the end of the Shang, as (the future) King Wu journeyed toward the capital. They were open to multiple interpretations. Flooding rivers and collapsing mountains pointed to cosmic irregularities and poor governance on the part of the Shang, but flooded roads and landslides could equally slow the path of an advancing army. The handle of the broom-star (the tail of a comet) pointed toward the Shang in the east, and might be read as a sign that the Zhou could be swept away. Yet this sign too was of uncertain significance. As another text put it, "When a broom-star appears, if our opponents control the handle's shaft, and we control its pointed tip, then they have the advantage in sweeping away, but we have the advantage in attacking" 彗星見, 彼操其柄, 我操其標, 以掃則彼利, 以擊則我利.[45] In the face of this apparently ambiguous series of signs, King Wu rallied his troops without threat of violence or offer of reward.[46] Even before all of his men could bloody their blades, the former subjects of Shang recognized his legitimacy. King Wu appears as an ideal reader of celestial signs who pushed onward in his campaign, having recognized that Heaven itself had turned against the wicked King Zhow. King Wu succeeded because he interpreted both the situation and the signs correctly and took the precise actions these demanded.[47] Assured of his own rightfulness, King Wu carried out the Punishment of Heaven.

Sign-Reading in the *Shiji*

While the *Huainanzi* integrated technical materials into an argument for governance aligned with cosmic patterns, the *Shiji* went further, constructing a contiguous historical narrative. The "Celestial Offices" treatise identifies court officials skilled in reading celestial signs—men like Sima Qian, his father, and his forefathers—as indispensable to good governance. Sima Qian, as Taishi ling 太史令 (Senior Archivist), served as a reader, recorder, and interpreter of omens past and present.[48] The *Shiji* "Basic Annals" describe ideal rulers and dynastic founders as sages able to accurately interpret the will of Heaven and submit to it. The poorest rulers ignore baleful signs, or worse still, fashion the signs of their own demise. Reading, interpreting, and responding to signs of all types proves a fundamental factor in explaining historical change in the *Shiji*.

Lineage and History in the "Celestial Offices"

The concluding essay in Sima Qian's "Celestial Offices" couches technical information on celestial signs within a broader discourse on effective governance. Sima emphasizes the value of the observation of the heavens to courts ancient and contemporary:

太史公曰: 自初生民以來, 世主曷嘗不曆日月星辰？及至五家、三代, 紹而明之, 內冠帶, 外夷狄, 分中國為十有二州, 仰則觀象於天, 俯則法類於地。天則有日月, 地則有陰陽。天有五星, 地有五行。天則有列宿, 地則有州域。三光者, 陰陽之精, 氣本在地, 而聖人統理之。

The Senior Archivist said: Since the very birth of the people, who among the lords of the various ages has not calculated the movements of the sun, moon, planets, and stars? With the advent of the Five Houses[49] and the Three Dynasties, the practice continued and was further illuminated. In the inner lands, there were those who wore caps and belts, and in far off places, there were the Yi and Di. The sages divided the central domains into twelve regions, and "gazing up, they observed the signs in the heavens, and lowering their heads, they emulated the categories of the earth."[50] In Heaven, there are the sun and moon, and on Earth, *yin* and *yang*. In Heaven, there are the Five Planets, and on Earth, the *wuxing* (Five Resources). In Heaven, there are the arrayed lodges, and on Earth, regions and borders. The Three Luminaries[51] are the essence of *yin* and *yang*. *Qi* originates in the Earth, and the sages unify it and put it to order.[52]

Sima Qian's essay declaims the historical significance and sagely character of the practices detailed in the body of the treatise. For Sima, celestial observation is part of a much grander enterprise, attributed to the acumen and perspicacity of the sage-kings of antiquity.

Sima Qian constructed a lineage of practitioners of the arts of celestial observation and astronomical calculation, a lineage to which he and his father, Sima Tan 司馬談 (Taishi ling, ca. 140–110),[53] belonged.[54] Beginning with its earliest practitioners in high antiquity, the lineage continued into the early Western Han. Sima Qian names three experts in celestial calculation (*tian shu* 天數) from his own dynasty. Tang Du 唐都 (fl. after 140 BCE) specialized in reading the stars, Wang Shuo 王朔 (fl. 110 BCE) in reading *qi*, and Wei Xian 魏鮮 (2nd cent. BCE) in reading signs related to the coming year.[55] Tang, Wang, and Wei serve both as conservative transmitters of sagely knowledge and as innovators who kept that knowledge relevant and useful in their own world. In compiling the "Celestial Offices," Sima Qian positioned himself as

Sign-Reading and Sagely Rulership in the *Huainanzi* and *Shiji* | 65

heir to this lineage, gathering together and organizing the sum total of knowledge needed to transmit the sagely enterprise. Sima's treatise not only encompassed the expertise of Tang, Wang, and Wei but surpassed it by juxtaposing political history with the appearance of celestial signs. He writes:

蓋略以春秋二百四十二年之間，月蝕三十六，彗星三見，宋襄公時星隕如雨。天子微，諸侯力政，五伯代興，更為主命。自是之後，眾暴寡，大并小。秦、楚、吳、越，夷狄也，為彊伯。田氏簒齊，三家分晉，並為戰國。爭於攻取，兵革更起，城邑數屠，因以饑饉疾疫焦苦，臣主共憂患，其察機祥候星氣尤急。

In the 242 years of the Chunqiu era, there were thirty-six lunar eclipses, three appearances of broom-stars, and in the time of Duke Xiang of Song (r. 650–637), stars fell like rain.[56] The Son of Heaven was diminished, and the local lords took power over the administration. The Five Hegemons[57] arose in turn, issuing their own commands in place of those of the ruler. From this time forth, the many did violence to the few, and the large absorbed the small. Qin, Chu, Wu, Yue—the Yi and Di—became powerful hegemons. The Tian ministerial clan usurped power in Qi.[58] Three Houses divided the domain of Jin.[59] All became Warring States, struggling for glory and gain, repeatedly raising the weapons of war, again and again slaughtering [the populations of] entire cities, inaugurating [an era of] starvation, pestilence, burning, and bitterness. Lords and ministers alike feared calamities, and thus they considered the examination of fortune and the observation of planets and *qi* to be a matter of great urgency.[60]

Sima portrays the observation of celestial signs as an art inextricably linked to the reading of political fortunes. The eclipses, appearances of comets, and meteors that "fell like rain" during the Chunqiu period accompanied the decline in power of the Zhou rulers and presaged the escalating violence and fragmentation of the Warring States period.

Sima notes that ancient celestial signs not cited in classical texts "cannot be studied in the present" 未有可考于今.[61] Such signs continued to appear during moments of political crisis in the Qin and Han. Following the appearance of numerous comets during the time of Qin Shihuang, "with the rise of the Han, the Five Planets gathered in Eastern Well" 漢之興, 五星聚于東井 in 205.[62] In the generations to follow, however, ill omens appeared. The treatise associates a solar eclipse with the rebellion of members of the clan of Empress Lü in 180 and a comet with the Seven Kingdoms Rebellion of 154.[63] Sima deployed vivid, macabre language to present the consequential relationship between the latter sign and the political events that followed it: "Heaven's Hound passed over the celestial field [corresponding to] Liang, and warfare arose. Soon thereafter, fallen corpses bled beneath it" 天狗過梁野; 及兵起, 遂伏尸流血其下.[64]

Sima Qian's essay underscores the value of interpreting celestial signs by pointing to their appearance in conjunction with pivotal historical events. Moreover, by constructing a tradition extending back to the sage-kings, the essay indicates the essential role of expert interpreters. Sima ends his treatise with a suggestion that technical knowledge of celestial signs can be fully realized only when paired with an investigation of historical events: "From the end back to the beginning, from antiquity up to the present, if we thoroughly examine the changes of the times, fathoming their finest points and crudest forms, then the 'Celestial Offices' will be made complete" 終始古今, 深觀時變, 察其精粗, 則天官備矣.[65] The "Celestial Offices" appeals to the authority of lineage and the strength of historical precedent.[66] Despite its similarities to the later "Celestial Patterns" in the *Hanshu*, it does not extensively employ frameworks drawn from the Classics nor direct citations of classical texts. It belongs to a time prior to the intimate interweaving of classical and technical knowledge that would occur in the late Western and Eastern Han.

In the "Celestial Offices," Sima Qian claimed that the observation of celestial signs was an enterprise within a sagely lineage that maintained its pragmatic political value in his own day. *Qi* rose up into the skies from the human realm. Reading celestial signs allowed the ruler to understand Heaven's response to developments within the empire itself. With the rise of Liu Bang, the Western Han founder, the Five Planets gathered in Eastern Well, while during the Lü clan rebellion an eclipse occurred, and in the months prior to the Seven Kingdoms Rebellion a comet appeared. Yet historical examples in the treatise itself are relatively sparse and primarily concern events in the Qin and Western Han. In the "Basic Annals," however, Sima Qian presented a powerful historical argument for the value of interpreting and responding well to all manner of signs.

Sign-Reading in Antiquity in the "Basic Annals"

The "Basic Annals" present a series of ideal rulers and ministers who prove to be effective readers of signs, perspicacious courtiers and monarchs who know whether those signs call for action or restraint. Reading these chronicles through the end of the Western Zhou reveals the ways in which sign-reading enters into dynastic change. The ideal ruler responds to Heaven, just as Heaven responds to him. The greatest of rulers, Yao, Shun, and King Wen, developed techniques and technologies that made signs comprehensible. Ideal rulers, often but not always dynastic founders, were vested with Heaven's Charge (*tian ming* 天命, a.k.a. the "Mandate of Heaven") and the responsibility for implementing Heaven's Punishment (*tian fa* 天罰). Yet decline always followed, though it was sometimes interrupted with a period of restoration under the odd good king, such as King Wuding 武丁 (r. 1250–1192) of Shang or King Xuan 宣 (r. 827–782) of Zhou. In the end, however, a ruler always came along who showed contempt for Heaven or complacency in his possession of its Charge (*ming* 命). No matter how clear the warning, the worst of rulers failed to yield to the will of Heaven.

Sima's first fleshed-out portrait of a cosmic king is that of Yao. Sima drew on the "Canon of Yao" (Yao dian 堯典), though his version differs from the received *Documents* text in some respects. Most notably, Sima introduces the sage-king through a series of similes not found in the *Documents* text: "His humanity was like that of Heaven, and his wisdom was like that of the gods. For those who drew near to him, he was like the sun. For those who gazed at him from afar, he was like a cloud" 其仁如天，其知如神。就之如日，望之如雲.[67] Sima's Yao shines forth, auspicious and awesome, benevolent and wise, a ruler who embodied the regularities of Heaven. Both the "Canon of Yao" and the *Shiji* passage describe him as a ruler who brings harmony to his own people and to the whole of humanity, noting that he drew others close, thereby expanding his influence, from the Nine Clans, to the Hundred Families, to the Myriad Domains (*wan guo* 萬國).[68] Yao himself was a visible, manifest sign of cosmic balance and regularity.

Yao created harmony in the world by modeling himself on cosmic patterns.[69] In both the *Shiji* and the *Documents*, he observed cosmic order in the heavens and applied it to his governance in the human world. Yao commanded members of the Xi 羲 and He 和 families to "take account of" or "enumerate" (*fa shu* 法數) the sun, moon, and stars in the *Shiji*, just as he commanded them to "trace the movements of celestial bodies" (*li xiang* 厤象)[70] in the "Canon of Yao." He sent Xi Zhong 羲仲 and Xi Shu 羲叔 to the east and the south, where they presided over the cosmic and mundane regularities associated with spring and summer. He sent He Zhong 和仲 and He Shu 和叔 to the west and north, where they attended to fall and winter. The Xi and He clans ensured that seasonal regularities continued unabated. Birds and beasts grew or shed their feathers or fur in a natural cycle that mirrored the human rhythms of agriculture. Under Yao, the ideal ruler, perfect order obtained throughout the cosmos. Long after his rule, however, the Xi and He grew corrupt under a poor ruler, and the cosmos became unbalanced and irregular.

Despite Yao's command of the cosmos, he never grew arrogant. He passed over his own son in favor of the self-sacrificing and deeply filial Shun. In his old age, he appointed Shun to administer in his stead, but delayed abdicating to him, taking time to "observe Heaven's Charge" (*guan tian ming* 觀天命).[71] When the moment came to select a successor, Yao remained circumspect and deferential to Heaven. Shun, duly attentive to celestial signs, proved up to the task: "Shun thereupon made inquiries concerning the *xuanji yuheng* (Agate Orb and Jade Crossbar), and so, regulated the Seven Administrators"[72] 舜乃在璿璣玉衡, 以齊七政.[73] Though the original sense of *xuanji yuheng* is unknown, in Han times, it came to be closely associated with celestial signs. Ma Rong and his student Zheng Xuan 鄭玄 (127–200) identified it as the armillary sphere, retrospectively casting Shun as a technologically adept observer of the heavens.[74]

In the *Shiji*, Heaven speaks through both signs in the firmament and the voices of ordinary people. Following the death of Yao, Shun offered to yield the throne to Yao's son, the wicked Dan Zhu 丹朱. It soon became clear, however, that Shun enjoyed the support of Yao's former subjects. Because of a series of human signs, he reasoned it was Heaven's will that he take the throne. When local lords ignored Dan Zhu and came to his court, asked him to judge their lawsuits, and sung his praises, Shun reluctantly said, "It is Heaven's doing" (*tian ye* 天也), and took the throne.[75] Yao's former subjects paid court to Shun, regarding him as a just arbitrator of disputes and a legitimate ruler. Despite his far-off wandering in the south and his respect for the tradition of passing power from father to son, Shun yielded to the will of the people. Just as much as signs in the sky, human actions pronounced the will of Heaven.[76]

The *Shiji* portrays King Wen of Zhou, like Shun, as an adept reader of signs whose legitimacy is underscored by the actions of human beings. In life, King Wen was known as the Earl of the West. While imprisoned by the last ruler of the Shang dynasty, King Wen created the Sixty-four Hexagrams of the *Changes* Classic, a sensitive instrument useful for identifying and interpreting subtle cosmic shifts. King Wen also proved a careful and reliable reader of human beings. Like Shun before him, King Wen was an authoritative and sought-after judge. Parties who brought their lawsuits to Shun and King Wen recognized their judgments as both binding and just.

Perhaps the most persuasive sign of King Wen's virtue was the influence he exerted upon those in his domain. The *Shiji* account of the lawsuit between Yu and Rui placed particular emphasis on the transformative power King Wen had on those who submitted to his rule, even when the last king of Shang was still in power. King Wen did not need to deliver an actual judgment; his influence was such that the parties settled their case out of court. Arriving in his domain, the men of Yu and Rui saw that farmers yielded the edges of the fields to one another and the young yielded to the old.

虞、芮之人未見西伯，皆慚，相謂曰：「吾所爭，周人所恥，何往為，祇取辱耳。」遂還，俱讓而去。諸侯聞之，曰「西伯蓋受命之君」。

Before the men of Yu and Rui had even had an audience with the Earl of the West, they were all stirred to shame and said to one another: "The men of Zhou would be ashamed to argue about the things that we do! How can we go there and do this? We will only humiliate ourselves!" Thereupon, they returned, yielding to one another as they went. When the local lords heard this, they said "The Earl of the West is the lord who has received the Charge."[77]

King Wen's transformative influence on the people of his domain, and anyone who entered it, signaled his legitimacy. It was the sign that he had received its Charge. His power was based on yielding rather than force. Yet when someone outside the current royal family received Heaven's Charge, with the rare exception of cases of abdication, the sitting ruler would be subjected to Heaven's Punishment. In the founding of the

Zhou dynasty, this grim responsibility would be left up to King Wen's son, the Martial King, King Wu.

In stark contrast to the story the *Huainanzi* told, in which the Zhou armies conquered the Shang capital without so much as drawing their weapons, King Wu could not avoid bloodying his own hands in the *Shiji* narrative. Nonetheless, King Wu in the *Shiji* was no less astute a reader of signs. He showed remarkable restraint in the face of his own bellicose generals, waiting until he was absolutely sure he had received Heaven's Charge before he dared to execute Heaven's Punishment.

Following his father's death, King Wu prepared for war. He promised the local lords who followed him that they would be rewarded for their achievements, and at the same time, his chief adviser, Lü Shangfu 呂尚父 (11th cent. BCE), promised that whoever lagged furthest behind would be beheaded (*zhan* 斬).[78] As they readied themselves to cross the Covenant Fords, the mood among King Wu and his ministers was warlike and grim. However, King Wu's understanding of what he had to do changed after he sacrificed a strange white fish that had leaped into his royal galley. When he roasted the fish, it transformed into a vermilion crow, and a fiery light descended from the heavens onto the Royal Chamber. The eight hundred local lords read these signs as an unambiguous message from Heaven that King Wu must strike the Shang ruler without delay: "The local lords all said, 'Zhow can be punished!' But King Wu responded, 'You do not understand Heaven's Charge. It is not yet permissible'" 諸侯皆曰:「紂可伐矣。」武王曰:「女未知天命, 未可也。」.[79] King Wu demonstrated himself to be a perceptive and cautious reader of signs. These omens meant not that he should go to war but that he should lie in wait. He admonished his men for their failure to understand Heaven's command and sent them home. Human signs—in the form of King Zhow's cruelty toward those most loyal to him—eventually led King Wu to carry out his campaign. Two years later, after King Zhow had murdered Bi Gan 比干, imprisoned Jizi 箕子, and drove his elder half-brother Weizi 微子 to flee to Zhou,[80] King Wu took an oath to "carry out the Punishment of Heaven" 行天罰.[81]

The "Basic Annals" tell the stories of not only the best but also the worst of rulers, knitting them together by lengthy genealogies, punctuated with brief periods of restoration. In the end, however, the histories of the Xia 夏 (trad. ca. 2070–ca. 1600), Shang, and Western Zhou are narratives of decline and dissolution. The worst rulers ignore or show contempt for both celestial signs and the welfare of their people.

In the Xia, problems arose almost immediately. Qi 啟, the son of the dynastic founder Yu 禹, ruled well. Yet despite the peaceful circumstances under which his father received the throne from Shun, he executed the Punishment of Heaven against the ruling clan of Hu (the You Hu Shi 有扈氏). Qi justified this campaign in a battlefield speech corresponding to the "Gan shi" 甘誓 (Oath at Gan) chapter of the *Documents*: "I swear before you: The clan in possession of Hu has imperiously polluted the *wuxing* and has neglected and abandoned the Three Rectifications. Heaven has employed me to destroy and cut off their charge. Now, I will carry out the Punishment of Heaven with you" 予誓告女: 有扈氏威侮五行, 怠棄三正, 天用勦絕其命。今予維共行天之罰。[82] Qi concluded his speech with a grim warning that those who failed to show bravery and discipline in battle would be slaughtered or enslaved along with their wives and children.[83]

As the Xia wore on, things got worse. The cosmic order Yao put in place eventually went awry, as the Xia produced less worthy rulers. A pivotal moment occurred during the reign of Di Zhongkang 帝中康: "The Xi and the He,[84] drunken and indulging in licentious behavior, abandoned [their responsibilities concerning] the seasons and let the days fall into disorder" 羲、和湎淫, 廢時亂日。[85] In an effort to set them right once again, Di Zhongkang sent his minister Yin 胤 to punish the Xi and the He.[86] A few generations later, local lords rebelled against Di Kongjia 帝孔甲 due to his licentious behavior and his fondness for spirits. His great-grandson, the Xia ruler Jie 桀, acted with such cruelty and violence toward his own subjects that he ultimately suffered the Punishment of Heaven at the hands of the Shang founder, Tang 湯.[87]

Sima Qian's account of the Shang dynasty is structurally similar to his account of the Xia. It begins with a story of justified conquest that soon devolves into a narrative of decline. In a speech much like Qi's "Oath at Gan," the Shang founder explained that he had not decided, of his own volition, to wage war. Heaven had commanded him to destroy Jie, and he dared not disobey:

匪台小子敢行舉亂, 有夏多罪, 予維聞女眾言, 夏氏有罪。予畏上帝, 不敢不正。今夏多罪, 天命殛之......爾尚及予一人致天之罰, 予其大理女。女毋不信, 朕不食言。

It is not that I, your Humble Son, would dare carry out a rebellion. He who possesses the Xia is guilty of many crimes.

> I have heard you all say that the House of Xia is rife with guilt. I am in terror of the High God, and I do not dare to fail to set things right. Now, the Xia has committed many crimes, and Heaven's Charge has been rescinded. . . . Come with me, the one man, to bring upon him the Punishment of Heaven, and I will greatly reward[88] you. Do not fail to believe me, for I will not eat my words![89]

Tang concluded his speech, as did Qi, by warning those who failed to show bravery that there would be dire consequences for themselves and their families. Such was the gravity of punishing Jie, a tyrant who threatened to rule as long as the sun reigned in sky. In the "Basic Annals" version of the speech, Tang alluded to a rhymed couplet attributed to Jie: "O sun, when at last shall you decay? When you and I both pass away" 是日何時喪？予與女皆亡！[90] Jie believed his power to be eternal. No sign, celestial or otherwise, could dissuade him. Only by force, Tang reasoned, could Jie be overcome.

Tang's victory over the Xia initiated a new dynasty, but that dynasty would have problems of its own. Tang's son, Taijia 太甲, began as a violent and cruel ruler, and the worthy minister Yi Yin 伊尹 forced him into exile. Three years later, once Taijia had reformed, Yi Yin yielded the reins of government to him. Within a few generations, the Shang began to decline. Periods of restoration occurred under three rulers: King[91] Taiwu 太戊, King Pan'geng 盤庚, and King Wuding. An omen in the time of King Wuding prompted the restoration that occurred during his reign.[92] When a pheasant lit on a cauldron during a sacrifice, Zuji 祖己[93] counseled Wuding on the best course of action. His speech made a powerful argument against the notion of Heaven as a capricious, willful entity whose commands and punishments are the products of mere whims. Zuji claimed that Heaven observes the human world, cutting short the lives of individuals and the lives of dynasties, based on what it sees.[94] Convinced, King Wuding paid heed to the signs, "cultivated his administration, put his virtue into action, brought joy to the world, and restored the Way of Yin" 修政行德，天下咸驩，殷道復興.[95] The story, also the subject of the "High Ancestor's Day of Sacrifice" (Gaozong rong ri 高宗肜日) chapter of the *Documents*, appeared in Han memorials that aimed to convince rulers to enact ritual reforms.[96] King Wuding's proper response to the pheasant on the cauldron made him a great ruler. But he was the last great ruler in the Shang.

Two generations later, under the violent and licentious King Jia 甲 (trad. d. 1148 BCE), the Shang again began to decline. King Wuyi 武乙 (trad. r. 1147–1113) was the worst ruler so far, engaging in vile acts of blasphemy against Heaven. Like Shun and King Wen, Wuyi was a creator of signs, but his creations were a grotesque sacrilege:

帝武乙無道，為偶人，謂之天神。與之博，令人為行。天神不勝，乃僇辱之。為革囊，盛血，卬而射之，命曰「射天」。

King Wuyi did not possess the Way. He constructed a puppet, which he called the Spirit of Heaven. He gambled with it, ordering men to make it move.[97] When the [so-called] Spirit of Heaven lost, he then butchered and humiliated it. Wuyi also constructed a leather bladder, which he filled with blood, raised up high, and shot with an arrow. He called this "Shooting Heaven."[98]

In stark contrast to careful readers of signs, King Wuyi had no fear of Heaven, but ritually murdered it. Even Jie, who likened himself to the sun, did not denigrate it. King Wuyi created the idol—an iconic representation of Heaven—for his own perverse amusement, and Heaven produced its own signs in response, striking him dead with a thunderbolt.[99]

The case of King Wuyi's great-grandson, King Zhow 紂 (r. 1075–1046) was mundane by comparison, but somehow more horrific. King Zhow was an arrogant yet clever young man, "intelligent enough to refute admonitions and eloquent enough to tart up lies" 知足以距諫，言足以飾非.[100] Always fond of women and wine, Zhow became truly vicious under the influence of his consort Daji 妲己, whom he obeyed without question. He indulged in licentious forms of music and raised taxes to build ever-more luxurious palaces and gardens. Along with Daji, he went to greater and greater excesses. "He constructed a lake of wine and a grove of trees, from which meat hung [like fruit], and ordered men and women to chase one another naked in it, while he drank the whole night through" 以酒為池，縣肉為林，使男女倮相逐其閒，為長夜之飲.[101] King Zhow fashioned a cosmos in miniature that reflected his unmoored desires. Like King Wuyi, King Zhow was a creator of perverse signs.

The signs of Zhow's incipient downfall were more human than celestial. When greater and greater numbers of local lords flocked to the Earl of the West, King Zhow's loyal servant, Zuyi 祖伊, warned the king that this was a sign of blame (*jiu* 咎). Zuyi's speech closely corresponds to the "The Earl of the West Conquers Li" (Xibo kan Li 西伯戡黎) chapter of the *Documents*.[102]

> 天既訖我殷命，假人元龜，無敢知吉，非先王不相我後人，維王淫虐用自絕，故天弁我，不有安食，不虞知天性，不迪率典。今我民罔不欲喪，曰『天曷不降威，大命胡不至』？今王其柰何？

> Heaven has rescinded its Charge from Our Yin. None among the readers of the great tortoise shells dares claim an auspicious result. It is not that the former kings do not succor us, their descendants, but rather that His Majesty has cut off the Charge by means of his violent and licentious behavior. Heaven has abandoned us, and we have neither a secure supply of food, nor means to predict Heaven's inclinations, nor to guide the people to follow the norms. Now, there is none among our subjects who does not wish for our demise. They ask, "Why does not Heaven send down a stern warning? Why does not [another] Great Charge arrive?" Now, Your Majesty, what are you going to do?[103]

King Zhow's reply echoed the Tyrant Jie's claim of ever-lasting authority: "Was I not born vested with the Charge from Heaven"[104] 我生不有命在天乎? Like Jie, King Zhow was certain he would always possess the charge, even as he destroyed any hope of recovering it. For King Wu, at least according to the *Shiji*, it was not the white fish or the vermilion bird that prompted him to execute Heaven's Punishment against Zhow, but rather the signs of Zhow's own fashioning. In murdering and mutilating the bodies of his most loyal ministers, Zhow created the signs that led King Wu to conclude that the Shang had truly lost its legitimacy. But the Zhou dynasty would no more last forever than had its predecessor.

Following the initial skirmishes in the first years of the dynasty, peace and prosperity reigned for more than a century, according to the *Shiji* narrative. The Zhou nonetheless declined, like the Shang and the Xia before it. It reached a low point under King Li 厲 (r. 877–841), a ruler profoundly uninterested in examining indicators of the effectiveness of his own administration. He suppressed criticism of his regime and conducted surveillance on his people, against the good advice of the

Co-Chancellor, the Duke of Shao 召公. Cruel and profligate, King Li left his subjects with no means of venting their anger against him. His allies among the local lords deserted him, and no one warned him when he was in danger. When open rebellion against the obnoxious king finally occurred, he had no means to defend his position, and consequently, he died in exile.[105] The crown prince, the future King Xuan, was left in the care of the Duke of Shao. When men of the capital surrounded his home, the Duke of Shao sacrificed his own son to the blood-thirsty mob to save the young prince. Raised under the tutelage of the Co-Chancellors, the Duke of Shao and the Duke of Zhou 周公 during their period of shared regency (r. 841–828),[106] King Xuan proved to be a good ruler, notwithstanding two major military defeats late in his reign. He relied extensively on the two former regents, invoking a renewal of the "remnant airs of King Wen, King Wu, King Cheng (r. 1042–1021), and King Kang (r. 1020–996)" 文、武、成、康之遺風.[107] For a few decades, the halcyon days of the early Western Zhou appeared to have returned. The reign of King Xuan's son, however, was an unmitigated disaster.

After his father's long reign, King You 幽 (r. 781–771) assumed power. Signs of Heaven's disfavor toward King You appeared early in his reign: "King Xuan's son, the Benighted King, Gongsheng, took the throne. In his second year, the Three Rivers in Western Zhou all suffered earthquakes" 子幽王宮湦立。幽王二年, 西周三川皆震.[108] The Court Diviner Boyang Fu 伯陽甫 explained the disturbance in yin and yang as a sign that the Zhou was coming to an end as the result of human actions, saying: "Zhou is about to perish. The qi of Heaven and Earth does not [spontaneously] lose its order. If its order goes astray, [it is because] the people have disordered it" 周將亡矣。夫天地之氣, 不失其序; 若過其序, 民亂之也.[109] Boyang described the Zhou as a domain where mountains were crumbling and rivers were blocked, recalling similar cosmological conditions in the last days of the Xia and the Shang. According to the "Basic Annals," his prophecy soon came true: "In that very year, the Three Rivers ran dry and Qi Mountain[110] crumbled" 是歲也, 三川竭, 岐山崩.[111] Human actions nonetheless work in conjunction with cosmological conditions. In the Shiji account, King You fell, like King Zhow of Shang, in part because he became enchanted with a femme fatale figure, Bao Si 褒姒, a stone-faced beauty who never laughed or smiled. He promoted her to Queen Consort, dismissing the daughter of the Marquis of Shen 申侯. In an attempt to make her laugh, King You lit signal fires to summon the local lords to defend against attackers. When they arrived and realized the king and queen had made fools of them, Bao Si finally laughed. King You lost the trust of his local lords, and when the Marquis of Shen, the Western Yi 西夷, and the Quan Rong 犬戎 laid siege to his court, no one came to his aid. The king was murdered. Bao Si was captured. The Zhou kings that followed held no more power than minor local lords. King You's reign left the Zhou dynasty a hollow shell, bringing an end to the Three Ages (sandai 三代) in all but name.

The "Basic Annals" supported Sima Qian's claims regarding the ancient status of the practice of interpreting celestial signs, yet made no distinction between celestial signs and other types of signs. An earthquake or a river changing course was no less damning to a sitting ruler than was a celestial body straying from its path. The "Basic Annals" contained a motley array of signs—strange birds, the white fish, earthquakes, and the perfect or imperfect movements of celestial bodies. Rulers who claimed to have attained Heaven's Charge, or who would carry out Heaven's Punishment, often cited human signs as authoritative proofs, for signs in the human world conveyed Heaven's will just as readily as signs in the heavens. Shun determined that Heaven had chosen him, rather than Yao's nefarious son Dan Zhu, when he found that various nobles came to his court, sung his praises, and trusted in his judgments. And when Zuyi warned Zhow that Yin has lost Heaven's Charge, he did so not on the basis of an obvious omen but rather on the basis of King Wen's conquest of Li.

Needless to say, the "Basic Annals" did not discount the importance of signs in the heavens. Yao's greatest achievement, matched only by his selection of Shun, lay in putting the heavens to order and emulating that order in his governance of the empire. Shun created new technologies, allowing more precise observation of celestial signs. King Wu was no less an able reader of Heaven's will than Yao or Shun, though he deciphered its intent in the white fish and vermilion bird, signs that would have no place in Sima Qian's "Celestial Offices." The "Basic Annals" lent sagely authority to the discourse surrounding celestial signs, though the chronicles concerned many types of signs. In terms of ontological quality and rhetorical value, celestial signs were not fundamentally different from other types of signs. They authorized those who could interpret them to make claims about the will of Heaven, just as other kinds of signs did. The difference between celestial signs and other kinds of signs consisted in the technical expertise needed to read them.[112]

Conclusion

The *Huainanzi* and the *Shiji* both position technical materials like those found in the Mawangdui tombs within massive textual projects concerning statecraft. Yet they do so in markedly different ways. The *Huainanzi* is a collection of twenty treatises, with the contribution each makes to statecraft defined in its twenty-first chapter. The organization of the *Huainanzi* is topical rather than chronological, thematic rather than historical. The *Shiji*, in contrast, examines change and continuity, positioning every event within historical patterns of revolution, decline, renewal, and dissolution. It embeds chronology even where its ostensible organization is thematic. Its treatises show how technical arts developed and were employed over time. The "Celestial Offices" concerns the persistence of a lineage of technical practitioners across dynasties and omens that mark historical change. The "Basic Annals" present an extended narrative in which omens announce the rise and fall of dynasties, sage readers divine the will of Heaven, and arrogant rulers ignore, or even create, the signs of their own demise.

Both the *Huainanzi* and the *Shiji* were assembled from fragments of other texts. The *Huainanzi* is representative of a place and a time in which the Five Classics had yet to obtain the unique status they would later acquire. While the *Huainanzi* does include some direct citations of the Classics, including ten marked references to the *Odes* and twelve to the *Changes*, the most important *jing* in the *Huainanzi* is the *Daodejing*, which the *Huainanzi* explicitly cites more than fifty times in its "Responses of the Way" (Dao ying 道應) chapter. Most of its sources, including the technical texts and numerous passages corresponding to parts of the received *Zhuangzi*, are not cited at all. It includes technical passages and illuminating narratives because of their inherent value rather than for the authority vested in them by virtue of their inclusion in a classical text.

The *Shiji* "Basic Annals" include many passages that closely correspond to the received *Documents*. The *Shiji* also appears to have drawn on other texts similar in form and content to *Documents* materials but not included in its received version. The *Shiji* employs the *Documents* primarily as a source of narratives surrounding tyrants and sage-kings rather than a source of classical proof-texts. Sima Qian marshals hallowed, sagely support to authorize a historical narrative in which acts of creating, reading, revering, denigrating, and ignoring signs prove pivotal to historical development. *Documents* materials are included less because they are part of an established Classic than because they tell the reader something about the best and worst rulers of the mythic and historical past. The "Celestial Offices" treatise likewise largely eschewed direct citation of the Classics, yet nonetheless appealed to sagely authority enshrined in a lineage traceable to exemplary antique models. In the late Western Han and Eastern Han, however, the Classics would become an increasingly important source of rhetorical authority in technical writing.

Chapter 3

Integrating Classical Texts with Omen-Reading Technology

In the late Western Han, classical texts took on an integral role in technical writing. Emperor Wu's expansionist policies had left the empire stretched thin, both financially and militarily. Reform-minded officials called for moving away from the Qin model of governance, advocating the restoration of the model of the early Western Zhou, as they understood it through textual traditions. In the decades following Emperor Wu's death, imperial princes studied the *Analects*,[1] advocates of retrenchment and austerity gained traction at court following the 81 BCE Salt and Iron Debates, and multiple traditions of classical learning flourished and were brought into dialogue at the Stone Canal Pavilion (Shiqu ge 石渠閣) conference of 51.[2] In 27, Emperor Cheng's court instituted a massive project to recover and catalogue texts from the collections of nobles and minor courts throughout the empire. Liu Xiang and Liu Xin 劉歆 (46 BCE–23 CE), father and son, spearheaded these efforts, cataloging and collating the collected works of the imperial library. Yang Xiong 揚雄 (53 BCE–18 CE) studied dialects to better understand ancient texts, elevated the Five Classics, and began composing new works on the models of the *Changes* and the *Analects*.

Liu Xiang and Liu Xin were experts in the Classics and omen reading. Ban Gu's 班固 (32–92) description of Liu Xiang's "Discourse on the Wuxing Tradition of the 'Great Plan'" (Hongfan Wuxing zhuan lun 洪範五行傳論) indicates the extent to which the technical art of sign-reading had become interwoven with the authority of the Classics in the late Western Han:

79

向見《尚書》《洪範》，箕子為武王陳五行陰陽休咎之應。向乃集合上古以來歷春秋六國至秦漢符瑞災異之記，推迹行事，連傳禍福，著其占驗，比類相從，各有條目，凡十一篇，號曰《洪範五行傳論》，奏之。

Liu Xiang saw that in the "Great Plan" chapter to the *Venerable Documents* Jizi had set forth the adulatory and condemnatory responses of *yin* and *yang* and the *wuxing* for King Wu of Zhou. Liu Xiang thereupon collected and collated a record of auspicious tallies and emblems, disasters and anomalies, from high antiquity through the Chunqiu and Warring States periods, up until the Qin and Western Han. He traced back actions and events, connecting the tradition to calamities and good fortune, noting both sign-readings and verifications. He arranged them by category in an organized fashion, and for each there is a descriptive heading. In total, it included eleven chapters (*pian*). He called this the "Discourse on the Wuxing Tradition of the 'Great Plan'" and submitted it to the throne.[3]

While it is uncertain to what extent Liu Xiang's text formed the basis for the *Hanshu* "Wuxing Treatise" (Wuxing zhi 五行志),[4] technical treatises in the *Hanshu* draw on the "Great Plan" and the *Annals* as essential sources of authority. The "Wuxing" surveys signs in the Chunqiu period, the Qin, and the Western Han dynasty, constructing a veritable sequel to the *Annals* Classic. Just as the *Gongyang zhuan* 公羊傳 (Gongyang Tradition) for the *Annals* proceeded toward Kongzi's famous declaration that his way was at an end,[5] the final version of the "Wuxing" developed teleologically toward the rise of Wang Mang 王莽 (r. 9 CE–23 CE) and the fall of the Western Han.

The task of disentangling Ban Gu's classicism from that of earlier figures is difficult, as significant uncertainty remains concerning who is responsible for which parts of the "Wuxing Treatise." As Nylan puts it, "All good scholars cannot but wonder whether Ban Gu imported another's views wholesale . . . revised the earlier sources he used beyond recognition, or wrote this treatise more or less from scratch."[6] To be sure, Ban's description of Liu Xiang's "Discourse on the Wuxing Tradition of the 'Great Plan'" does suggest similarities to the *Hanshu* "Wuxing Treatise." Just as Liu's text is divided into eleven chapters, the treatise consists in eleven sections, all derived from the "Great Plan." Each of these bears a descriptive heading, including a citation of a tradition and its explication, followed by records of "auspicious tallies and emblems, disasters and anomalies," along with "sign-readings and verifications." Like Liu's text, the treatise is "arranged . . . by category in an organized fashion," and it is likely based, at least to an extent, on his work. If this is indeed the case, there is evidence of both abridgment and expansion, as the treatise includes precious few signs from high-antiquity and the Warring States period as well as a number of signs that postdate Liu's death. While these may have been added by Liu Xin, they are presented in a manner that condemns Wang Mang, and would have been considered treasonous during his reign. Of all parts of the treatise, its framing introduction and those entries that specifically warn of the rise of Wang Mang appear to bear the deepest imprint of an Eastern Han compiler, presumably Ban Gu.

Ban Zhao 班昭 (ca. 48–ca. 116), Ma Rong 馬融 (79–166), and Ma Xu's 馬續 (ca. 79–after 141) "Treatise on Celestial Patterns" (Tianwen zhi 天文志),[7] compiled some years following Ban Gu's death, is a natural point of comparison with Sima Qian's "Celestial Offices." The "Celestial Patterns" adopts the basic structure of the "Celestial Offices" with much of its text consisting in materials derived verbatim from the earlier work. The differences between the two, however, illuminate the elevated role classical authority had come to play in technical writing by circa 100 CE. The "Celestial Patterns" places greater responsibility on the emperor and his court for the appearance of inauspicious omens, liberally employs quotations from the Classics as proof-texts in new technical passages, and includes a teleological set of annals that punctuates the story of the fall of the Western Han with a series of baleful celestial signs.[8] Despite its highly visible similarities to the "Celestial Offices," the "Celestial Patterns" shares its historiographical purpose and basis in classical authority with the "Wuxing."

The "Wuxing Treatise" and Its Classical Models

The "Wuxing," in contrast to the "Celestial Offices," interweaves celestial signs and other omens into an interpretive framework derived from the Classics. The introduction to the "Wuxing" positions it as both an extension of the "Great Plan" chapter of the *Documents* and a Han expansion of the *Annals*. The "Wuxing" records comets and eclipses in greater numbers than even the "Celestial Patterns." It concerns the postmortem historiographical analysis of the fall of a dynasty—omenology writ large.

The "Wuxing" claims that signs are inherent in the cosmos, but made readable by sage figures who produce and extend the Classics. Its introduction interweaves the Nine Divisions of the "Great Plan" with the Eight Trigrams of the *Changes*. According to the treatise, sages had produced these as textual emulations of the Yellow River Chart and the Luo Writings, signs Heaven itself had wrought. The Trigrams and the Divisions made up the core of the *Changes* and the "Great Plan." Like the Classics and their commentaries, they were bound together as "warp and weft" (*jing wei* 經緯). Later sages, such as King Wen and Kongzi, extended and elaborated upon them during moments of historic decline. King Wen expanded the *Changes* by producing the Sixty-four Hexagrams in the last generation of the Shang. Kongzi compiled the *Annals*, recording events that verified the "Great Plan."[9] The introduction to the "Wuxing" tells a story in which the creation of the Sixty-four Hexagrams and the composition of the *Annals* constitute sagely responses to desperate times. King Wen and Kongzi lived in times when the will of Heaven had been ignored or forgotten, yet they produced texts that could be used to recover and re-enact that will.

Compiled nearly a century after the collapse of the Western Han, the *Hanshu* places the "Wuxing" within the lineage of the *Annals*, and by extension, within the lineage of the "Great Plan" and the Luo Writings. The treatise presents the views of a range of readers of omens, almost all active during the late Western Han, who had read contemporary signs by comparing them to classical precedents. Dong Zhongshu 董仲舒 (ca. 195–ca. 104)[10] was the single exception to this tendency, though his views are often conflated with those of Liu Xiang. Liu Xiang and Liu Xin were most frequently cited, with 120 and 80 respective entries explicitly including their views.[11] The compiler concludes: "I have gathered together . . . the explanations of various court affairs up until the time of Wang Mang, for a total of twelve generations, to serve as a supplement to the *Annals*, and compiled this chapter" 是以攬 . . . . . . 所陳行事, 訖於王莽, 舉十二世, 以傅《春秋》, 著於篇.[12] The treatise is fashioned as the culmination of Western Han omenology and an extension of the Classics.

The "Wuxing" is at once an expansive set of commentaries to the "Great Plan" and an annotated *Annals* for Han times. It employs and distinguishes between the interpretations of omens, during both the Chunqiu period and the Western Han, from the most prominent omenologists and experts on the *Annals*, the *Documents*, and the *Changes*. The macrostructure of the "Wuxing" is derived from three of the Nine Divisions in the "Great Plan" (see box 3.1): the *wuxing* (Five Resources/Phases), the Five Duties (*wushi* 五事), and the Sovereign's Standard (*huangji*

**Box 3.1
Organization of the "Wuxing Treatise"**

Preface 序 (*Hanshu* 27A)

Five Resources/Phases 五行

Wood 木
Fire 火
Soil 土
Metal 金
Water 水

Five Duties 五事 (*Hanshu* 27B)

BEARING 貌

Bearing not respectful 貌不恭
Excessive rain 恆雨
Clothing monstrosities 服妖
Calamities involving chicken 雞禍
Blue pestilences and omens 青眚、青祥
Metal disrupts wood 金沴木

SPEECH 言

Speech not yielding 言不從
Excessive sunshine 恆陽
Rhymed monstrosities 詩妖
Furry-beast banes 毛蟲之孽
Calamities involving dogs 犬禍
White omens 白祥
Wood disrupts metal 木沴金

Vision 視 (Hanshu 27C)

Excessive heat 恆奧
Grassy monstrosities 草妖
Feathered-beast banes 羽蟲之孽
Calamities involving sheep 羊禍
Red omens 赤祥

Listening 聽

Excessive cold 恆寒
Drumming monstrosities 鼓妖
Fish banes 魚孽
Scaly or shelled-beast banes 介蟲之孽
Calamities involving pigs 豕禍
Fire disrupts water 火沴水

Thought 思 (Hanshu 27D)

Excessive wind 恆風
Fatty (?) night monstrosities 脂夜妖
Naked-beast banes 蠃蟲之孽
Calamities involving cattle 牛禍
Ailments of the heart and stomach 心腹痾
Yellow omens 黃祥
Metal, wood, fire, and water disrupt soil 金木火水沴土

Sovereign Standard 皇極

Excessive darkness 恆陰
Archery monstrosities 射妖
Dragon and snake banes 龍蛇之孽
Calamities involving horses 馬禍
Ailments causing those below to attack those above 下人伐上之痾
The sun and moon move chaotically 日月亂行 (Hanshu 27E)
The planets move retrograde 星辰逆行

皇極).¹³ The materials included in these major sections might be read as commentary on the "Great Plan" itself. The first major subsections place omens in categories associated with each of the *wuxing*: wood (*mu* 木), fire (*huo* 火), soil (*tu* 土), metal (*jin* 金), and water (*shui* 水). The second major section categorizes omens under each of the Five Duties: bearing (*mao* 貌), speech (*yan* 言), vision (*shi* 視), listening (*ting* 聽), and thought (*si* 思). The final major section includes omens related to the Sovereign's Standard. Each category is further subdivided into specific kinds of disasters and anomalies that issue from imbalances in the *wuxing*, problems related to the Five Duties, or imperfect adherence to the Sovereign's Standard. Some omens fit into their categories according to a transparent logic, while others have opaque relationships to the categories in which they are organized. The Duty of Speech category, for instance, includes such likely suspects as "speech that is not yielding" (*yan bu cong* 言不從) and "rhymed monstrosities" (*shi yao* 詩妖) but also includes strange occurrences that bear no obvious relationship to speech, such as "excessive sunshine" (*heng yang* 恆陽), "calamities involving dogs" (*quan huo* 犬禍), and "white omens" (*bai xiang* 白祥).

The treatise uses the motley assortment of signs associated with the *wuxing*, the Five Duties, and the Sovereign's Standard as the basis for its organization, introducing each subsection with a quotation from a tradition of the "Great Plan"¹⁵ that details the sort of problems associated with the category under discussion. Like the *Annals*, the treatise organizes individual types of signs chronologically. First, the "Wuxing" lists entries in the *Annals* that record appearances of a given type of sign and Western Han interpretations. Second, the treatise lists instances of appearances of the same kinds of signs in the Qin and Western Han, likewise with Western Han interpretations. The "Wuxing" formally mirrors the *Annals*.

Celestial signs, along with certain human signs, such as people growing horns or spontaneously changing sexes, occur almost exclusively in the final major division of the "Wuxing," the section on the Sovereign's Standard:

傳曰:「皇之不極,是謂不建,厥咎眊,厥罰恆陰,厥極弱。時則有射妖,時則有龍蛇之孽,時則有馬禍,時則有下人伐上之痾,時則有日月亂行,星辰逆行。」

The tradition says, "When the sovereign does not attain the standard, this is called failing to establish it. Its sign of blame is the loss of vision, its penalty excessive darkness (*yin*), and its extreme manifestation is weakness. At times there are archery monstrosities, at times dragon and snake banes, at times calamities involving horses, at times ailments [leading] the people below to attack those above them, and at times the sun and moon move chaotically, or the planets move in retrograde against the stars.[16]

The lengthy explanatory section that follows makes the appearance of celestial signs, and other signs included in the same section, the product of lapses in all of the Five Duties. Taken as a group, these lapses constitute a failure to attain the standard. Cloudy *qi* rises from the mountains, covering the heavens, so that the "*qi* of Heaven is disordered" 天氣亂. The ruler is like the arrogant high-flying dragon in the line statements to the first hexagram in the *Changes*, Qian 乾 (Pure Yang).[17] The *Changes* says "clouds follow dragons" 雲從龍,[18] and so dragon and snake banes are born. The hexagram Qian corresponds to both rulers and horses, and so calamities involving horses occur. The explanatory section proclaims that the tradition, on analogy with the *Annals*, conceals a powerful critique of the ruler:

凡君道傷者病天氣,不言五行沴天,而曰「日月亂行,星辰逆行」者,為若下不敢沴天,猶《春秋》曰「王師敗績于貿戎」,不言敗之者,以自敗為文,尊尊之意也。

When the way of the ruler comes to harm, sickening the *qi* of Heaven, the text does not say that the *wuxing* have disrupted Heaven, but instead states that "the sun and moon move chaotically," or "the planets move in retrograde against the stars." This is much like the *Annals* stating that "the king's forces suffered defeat at Maorong."[19] It does not say who defeated him, but uses language suggesting that he was defeated of his own accord. The meaning of doing so is to exalt those who are worthy of exaltation.[20]

The "Wuxing" places responsibility for the appearance of celestial signs squarely on the head of the ruler. While it claims that the tradition avoided, out of ritual respect, any indication of the ruler as the source of disorder in the cosmos, it suggests that disorder in the heavens issues from disorder at court. The idea that celestial signs are rooted in ongoing processes in the human world colors the interpretation of such signs both in the *Annals* and in records from the Western Han.

The "Wuxing Treatise" Chronicle of Comets

The "Wuxing" repeatedly tells two stories. One is the story of the Chunqiu period, the slow descent toward the chaos of the Warring States period. The second has a somewhat different arc, beginning with Liu Bang's triumph over his rival Xiang Yu 項羽 (232–202) in the years of disorder that followed the collapse of the Qin dynasty. Yet, the second story, which always ends with the rise of the usurper Wang Mang, is likewise a narrative of decline. Almost any chronicle of almost any category of inauspicious sign can serve in the "Wuxing" as a chronicle of the decline and fall of the Western Han. The set of records pertaining to comets reflects broader formal features of the chronicles, presenting a typical, if elaborate, example of the narrative arc in which the "Wuxing" presents various signs in the Chunqiu and Qin-Western Han periods. Within these chronicles, the "Wuxing" proves to be a true supplement to the *Annals*, an *Annals* for Han times.

The *Annals* for the Chunqiu period recorded only three comets. The Han chronicle of comets includes detailed readings of these records attributed to Dong Zhongshu, Liu Xiang, and Liu Xin, as well as relevant passages from the *Zuozhuan*. Liu Xiang's commentary also cites the *Xingzhuan* 星傳 (Tradition of Stars and Planets)[21] several times. In the *Annals* Classic, as it is read in the "Wuxing," comets appear at crucial moments. A comet sweeps through Northern Dipper in 613 BCE, presaging the murders of the lords of Qi, Song, Lu 魯, Ju 莒, and Jin 晉 over the next twenty-eight years. A comet near the Great Star 大辰 (Antares) prompts one reader of celestial signs, whose words are reproduced from the *Zuozhuan*, to make a memorable and often repeated claim: "Broom-stars are the means by which the old is swept away[22] and the new is spread forth" 彗所以除舊布新也.[23] Many once glorious houses were indeed swept away following the final comet in the *Annals* (482 BCE). The "Wuxing" records, but does not differentiate between, the interpretations of Dong Zhongshu and Liu Xiang: "Dong and Liu thought as follows: The reason it does not say the name of the lodge is because it was not imposed on a lodge. When an asterism rises following the sun, disorderly *qi* occludes the brightness of the lord. The next year, the events in the *Annals* came to a conclusion" 董仲舒、劉向以為不言宿名者，不加宿也。以辰乘日而出，亂氣蔽君明也。明年，《春秋》事終。[24] The final comet appeared to herald the end of the Chunqiu period.

The chronicle of comets during the Western Han presents an arc that might be heuristically divided into five stages: the founding of the dynasty; early rebellions; military conquests; strife at court; and the rise of Wang Mang. The first stage, which includes a single comet appearing in 204 BCE, presages Xiang Yu's defeat at the hands of Liu Bang, the first Han emperor:

有星孛于大角，旬餘乃入。劉向以為是時項羽為楚王，伯諸侯，而漢已定三秦，與羽相距滎陽，天下歸心於漢，楚將滅，故彗除王位也。一曰，項羽阬秦卒，燒宮室，弒義帝，亂王位，故彗加之也。

A comet appeared near Great Horn,[25] and only after more than ten days did it set. Liu Xiang thought [as follows]: At that time Xiang Yu (232–202) was the King of Chu, and was hegemon over[26] the local lords, but the Han had already pacified the three districts of Qin,[27] and was distant from Xiang Yu, in the city of Xingyang. But the hearts of the whole world gravitated to the Han, and Chu was about to be destroyed. Thus, the broom-star cleared the throne. One source says: "Xiang Yu massacred [surrendered] Qin soldiers,[28] burned palaces, murdered Yidi (r. 208–205), and brought disorder to the throne. Thus, the broom-star was imposed upon him."[29]

This comet proves a baleful omen, but not for Liu Bang. A series of bad omens can be auspicious indeed for one who is bent on toppling a sitting ruler. The "Wuxing zhi" gives two possible interpretations, both inauspicious for Xiang Yu. Liu Xiang's interpretation identifies Xiang Yu as a king. After briefly occupying the throne, Xiang Yu is swept away. In the second interpretation, from an unknown Western Han source, Xiang Yu brings disaster on himself through a series of atrocities culminating in the murder of the previous King of Chu, Yidi (the Dutiful Emperor), whom Xiang Yu himself had enthroned only a few years earlier.

Comets in the second stage mark rebellions. The first occurred in 157, three years prior to the Seven Kingdom's Rebellion, and the second occurred in 135, thirteen years prior to the rebellion of Liu An, the King of Huainan. Liu Xiang's interpretation of the 157 BCE comet, which had appeared soon after the death of Emperor Wen 文 (r. 180–157), points to both the coming rebellion and Emperor Jing's triumph over it.[30] The historiography surrounding the second comet in this stage blends intrigue at court with rebellion. The treatise ties the comet to both Liu An's visit to court in 134 and the "haughty and self-indulgent" (*jiao zi* 驕恣) behavior of Empress Chen 陳皇后 (fl. 130 BCE).[31] Emperor Wu met both challenges, deposing Empress Chen and executing Liu An.[32] Comets announce rebellion and, in the case of Empress Chen, potentially destructive influences within the palace walls. The second stage emphasizes the capacity of early Western Han rulers to effectively respond to challenges.

The first comet of the third stage, identified as "The Banner of Chiyou" 蚩尤旗,[33] occurs only two months after the last comet of the previous stage. Stretching across the sky, the comet elicited an anonymous prognostication announcing: "the king will march in conquest on the Four Directions" 王者征伐四方. It appeared to herald the expansionist policies of Emperor Wu. An interpretation of the second comet in this stage, also without an identified source, points to the intensification of military campaigns in the north: "In the fourth month of the fourth year of the Yuanshou reign period (119 BCE), a long star again emerged in the northwest. At this time, the campaigns against the Hu grew especially extreme" 元狩四年四月, 長星又出西北, 是時伐胡尤甚.[34] Whereas the prognostication accompanying the first comet supported an expansionist campaign, the interpretation of the second points to the excessive and potentially dangerous point the expansionist policies had reached. The second comment is not a prognostication but a retrospective, critical judgment, representative of historical perspective. It works through a simple act of juxtaposition, artlessly bonding the trajectory of the history of the Western Han to the appearance of celestial signs.

The fourth stage marks in earnest the decline of the Western Han, but that decline is punctuated by a restoration under Emperor Xuan 宣 (r. 74–48). The three comets in this stage all accompany strife at court. The first occurs in 110 BCE, presaging the disastrous fighting near the end of Emperor Wu's reign, during which the Crown Prince would be executed and Empress Wei 衛 (d. 91 BCE) condemned for witchcraft. The second appeared during the last year of the life of Huo Guang 霍光 (d. 68 BCE), the *de facto* ruler of the empire since Emperor Wu's death. It anticipates Huo's demise and the destruction of his family two years later.[35] Huo's rule marks a period in which power did not rest in the hands of a Liu clan emperor. Like Wang Mang, Huo could not be a legitimate ruler, though unlike Wang, Huo never declared himself emperor. With Huo's death, legitimate rulers returned to power. Emperor Xuan's personal rule ushered in a mid-dynastic renaissance; for three generations the line of succession would be uninterrupted. But with the coronation of his grandson, Emperor Cheng, the process of decline began again. A 32 BCE comet in the lodge House[36] would be read as a sign that no heir would survive him. Both Liu Xiang and Gu Yong 谷永 (fl. 36–ca. 9 BCE) interpreted the comet as a sign that a pregnant woman would be murdered, cutting off the line of succession. Emperor Cheng would, in the end, nominate a nephew as his heir.[37] And while Emperor Ai 哀 (r. 7–1 BCE) proved no friend to Wang Mang, his early death without issue set the stage for Wang to become regent, and ultimately, to take the throne.

The final stage in the chronicle consists of a single, lengthy entry narrating an appearance of Halley's Comet[38] near the end of Emperor Cheng's reign. The entry includes three main components: a description of the spectacular trajectory of the celestial sign, the views of Gu Yong and Liu Xiang, and finally, a description of the events the comet portended up to the rise of Wang Mang. Gu identified its appearance with treachery in the rear palace and widespread revolt, while Liu compared it to signs that accompanied the fall of the previous dynasty. Ban Gu summarizes the events between the appearance of the comet and the fall of the dynasty, including the murder of two imperial princes, the suicides of two empresses, and the deaths of three emperors in eleven years,[39] after which Wang Mang seized power. The chronicle of comets ends, as do the chronicles of many types of signs in the "Wuxing," with the fall of the Western Han.

Etiology, Chronology, and Classicism in the "Celestial Patterns"

Ban Zhao and the Ma brothers synthesized aspects of the "Celestial Offices" with aspects of the "Wuxing Treatise" in their "Treatise on Celestial Patterns." They appropriated much of Sima Qian's treatise wholesale; large parts of the "Celestial Patterns" correspond very closely to the "Celestial Offices," with sections on the Five Planets, the sun, and the moon, incorporating the bulk of their new technical materials (see table 3.1). However, the "Celestial Patterns" distinguished itself from

Table 3.1. Comparison of *Hanshu* "Celestial Patterns" and *Shiji* "Celestial Offices"

Shiji "Celestial Offices"	Hanshu "Celestial Patterns"	Comments
	Introduction	No correspondence with *Shiji*
The Stars Central Palace Eastern Palace Southern Palace Western Palace Northern Palace	*The Stars* Central Palace Eastern Palace Southern Palace Western Palace Northern Palace	Close correspondence throughout
The Five Planets Jupiter Jupiter cycle Mars Saturn Planetary interactions Venus Mercury — —	*The Five Planets* Jupiter — Mars — — Venus Mercury Saturn Planetary interactions	The *Hanshu* adds new correspondences, including the Five Constants and the Five Modes of the "Great Plan," and re-orders the planets so that Saturn, associated with the Center and with Virtue (*de*), appears in the final position. Correspondence between treatises is limited.
—	Regional field allocations	Closely corresponds to section in *Shiji*
—	Sixty-day cycle, regional correspondence	Partial correspondence with *Shiji*, Ten-day cycle, regional correspondences
—	Planets and regions	—
—	Jupiter cycle	Cites "Shi shi" and "Gan shi"
—	Retrograde motion and eclipses	Cites Classics to argue against claims that eclipses and retrograde motion can be considered regular behavior

continued on next page

Table 3.1. Continued.

Shiji "Celestial Offices"	Hanshu "Celestial Patterns"	Comments
Regional field allocations	—	Shiji section contains one additional line
Sun and Moon	—	Little correspondence with cognate section in Hanshu, significantly shorter
Ten-day cycle, regional correspondences	—	Most of this section is also found in Hanshu, Sixty-day cycle, regional correspondences.
Emergent "Stars"	Emergent "Stars"	Primarily designations applied to comets and novae, with close correspondence throughout
—	Sun and Moon	More developed than cognate Shiji section, with several citations of Classics
Clouds and *qi*	Clouds and *qi*	Close correspondence throughout
Tetrasyllabic poem	Tetrasyllabic poem	Poem describes various baleful signs, including earthquakes, collapsing mountains, crop failures, and floods. Close correspondence throughout
Divining the harvest	Divining the harvest	Methods include examining weather, wind direction, tone emitted from populace, and the location of Taisui (a hypothetical planet with a position opposite Jupiter) at the beginning of the year. Close correspondence throughout

Integrating Classical Texts with Omen-Reading Technology | 93

Shiji "Celestial Offices"	Hanshu "Celestial Patterns"	Comments
Sima Qian's postface	—	Parts of Sima Qian's postface are incorporated into the *Hanshu* treatise, especially its concluding chronicle.
—	Chronicle through Aidi (7–1 BCE)	—

Sima's treatise in three important ways. First, the introduction to the "Celestial Patterns" replaces Sima's closing essay. Where Sima's comments framed technical materials as the patrimony of a lineage of experts, the opening of the "Celestial Patterns" emphasizes the etiology of celestial signs, declaring them to be direct reflections of events in the human realm. Second, the "Celestial Patterns" draws directly and explicitly on classical authority. It includes several technical passages not found in the *Shiji* treatise that cite the Classics as proof-texts. The classicism of the "Celestial Patterns," moreover, plays a vital role in the network of taxonomic correspondences within which it positions planetary signs. In contrast to Sima's treatise, it ties each of the Five Planets to one of the Five Duties in the "Great Plan" and to the "Monthly Ordinances" (Yue ling 月令) of the *Record of Rites* (Liji 禮記).[40] Finally, the "Celestial Patterns" closes with a chronicle of celestial signs that tells the story of the rise and fall of the Western Han.

Where the *Shiji* treatise speaks of the pragmatic value of celestial observation to the court, linking it to the traditions of high antiquity, the "Celestial Patterns" interprets celestial aberrations as signs that the ruler has endangered himself and his court due to his own ritual failures. Before launching into a description of the constellations of the Five Palaces nearly identical to the first section of the "Celestial Offices," the "Celestial Patterns" introduces the treatise with a declaration of the resonance between such celestial signs and the court:

其本在地，而上發于天者也。政失於此，則變見於彼，猶景之象形，鄉之應聲。是以明君覩之而寤，飭身正事，思其咎謝，則禍除而福至，自然之符也。

> [Portents] are what originate in the earth and erupt above into the heavens. When governance fails (down) here, then aberrations appear (up) there, just as shadows are the counterparts of their forms and echoes are responses to sounds. This is why the clear-sighted ruler sees them and awakens, putting himself in order and rectifying his affairs. Should he set his mind to repenting his crimes, then calamity can be avoided and blessing brought forth. This is the tally that comes about on its own.[41]

While the "Celestial Patterns" contains much of the same information as the "Celestial Offices," it frames that information differently. Sima's treatise moves directly into the technical details of which constellations are called what and located where. The "Celestial Patterns" contains the same information but prefaces it by claiming that the ruler and his court are responsible for the appearance of celestial signs. Inauspicious signs result from faults in governance or in ritual. Actions can be taken to avoid the calamitous outcomes associated with signs that have already appeared; prognosis is not to be confused with fate. The same technical information, presented in another way, might seem inevitable. The "Celestial Patterns" emphasizes that bad fortune can be turned into good. Auspicious signs signal that the court has earned the favor of the heavens.

The "Celestial Patterns" tells much the same story as the "Celestial Offices"; the passages it includes on Chunqiu-era portents and the political events that accompany them correspond closely to those in the *Shiji*. The "Celestial Patterns" recounts many of the same events in the Qin and early Han. The appearance of celestial signs as evidence that Heaven vested authority in the ruler is, however, articulated far more explicitly in the "Celestial Patterns" than in the "Celestial Offices." Where the "Celestial Offices" reports the conjunction of the Five Planets that accompanied the "rise of the Han,"[42] the "Celestial Patterns" alone refers to this sign as the "Tally that the Founding Emperor had received the charge" 高皇帝受命之符也.[43] The "Celestial Offices" argued for the pragmatic value of the sagely enterprise of reading celestial signs to the court. The "Celestial Patterns" claimed that celestial signs could indicate that a particular individual had received the right to rule from the heavens, but more often, celestial signs indicated ritual or administrative failure.[44]

Linking Signs to the Classics in the "Celestial Patterns"

While the *Shiji* "Celestial Offices" treatise had no doubt drawn from a variety of textual materials, it tended to efface rather than highlight its relationship to its textual antecedents. Despite the high level of correspondence between the main body of the *Shiji* and *Hanshu* treatises, the "Celestial Patterns" contains far more numerous references to the Classics than does its predecessor. Moreover, as is the case in the "Wuxing," the tradition for the "Great Plan" shapes the manner in which planetary omens are interpreted, linking them to specific types of ritual or political failures.

Whereas Sima Qian had claimed a sagely lineage for the practice of reading celestial signs, the "Celestial Patterns" vested the reading of celestial signs with the authority of the Classics. One technical passage in the "Celestial Patterns," but not the "Celestial Offices," integrated the movements of the moon into a spatial schema built on the *Changes* and bolstered by citations from the *Odes* and the *Documents*:

《易》曰「東北喪朋」。及《巽》在東南, 為風; 風, 陽中之陰, 大臣之象也, 其星, 軫也。月去中道, 移而東北入箕, 若東南入軫, 則多風。西方為雨; 雨, 少陰之位也。月去中道, 移而西入畢, 則多雨。故《詩》云「月離于畢, 俾滂沱矣」, 言多雨也。《星傳》曰「月入畢則將相有以家犯罪者」, 言陰盛也。《書》曰「星有好風, 星有好雨, 月之從星, 則以風雨」, 言失中道而東西也。

The *Changes* says: "In the northeast a friend is lost."[45] The trigram *Yielding* lies in the southeast and is wind. Wind is *yin* within *yang* and is the counterpart of the Great Minister. Its stars are those in Axletree. When the moon deviates from the middle path [i.e., the ecliptic], moves northeast and enters Basket, or if it moves southeast and enters Axletree, then there is much wind. The west is rain. Rain is the position of lesser *yin*. When the moon deviates from the middle path and moves west to enter Net, then there is much rain.[46] Thus when the *Odes* says, "The moon is caught in Net! How it makes the waters run,"[47] it speaks of there being much rain. And when the *Tradition of the Stars and Planets* says, "When the moon enters Net, there will be those among the generals and great ministers that commit crimes on account of their households," it speaks of the overabundance of *yin*. When the *Documents* says, "In the stars there is fine wind; in the stars there is fine rain . . . when the moon follows the stars, there is wind and rain,"[48] it speaks of [the moon's] deviation from the middle path as it moves from east to west.[49]

By citing the *Changes*, the *Documents*, and the *Odes*, the "Celestial Patterns" reinterprets the Classics themselves. Fragmentary though these citations may be, their use in this context implies that the Classics speak directly to such issues. The discourse of celestial signs, cobbled together from sundry technical texts, the "Celestial Offices," and only a few other named traditions, thus obtains classical authority.

The "Celestial Offices" and the "Celestial Patterns" contrast sharply in their treatment of the planets, and this contrast issues in part from the classicism of the *Hanshu* treatise (compare tables 2.2 and 3.2). The "Celestial Patterns" introduces several planetary correspondences, linked specifically to the Classics, that do not appear in the "Celestial Offices." In addition to the familiar correspondences with direction, season, and material, the "Celestial Patterns" ties each of the Five Planets to one of the Five Constants (*wu chang* 五常) and to one of the Five Duties. The emperor's failure in these duties, as the "Wuxing" argues, is liable to produce all manner of baleful signs. The "Celestial Patterns" associates planetary aberrations with failure to adhere to the ritual, economic, and military prescriptions of the "Monthly Ordinances" of the *Record of Rites* and identifies specific actions that the ruler and his court should take to maintain cosmic balance. The case of the Dazzling Deluder, Mars, presents a typical example:

熒惑曰南方夏火, 禮也, 視也。禮虧視失, 逆夏令, 傷火氣, 罰見熒惑。

The Dazzling Deluder is said [to correspond to] "the south, summer, and fire."[50] It is the rites and the (Duty of) Vision. When the rites diminish and there are failures in the (Duty of) Vision, or when the Summer Ordinances are violated, this damages the fire *qi*, and the penalty appears in the Dazzling Deluder.[51]

Table 3.2. Planetary Correspondences in the *Hanshu* "Celestial Patterns"

	Jupiter	Mars	Venus	Mercury	Saturn
Direction	East	South	West	North	Center
Season	Spring	Summer	Autumn	Winter	Late Summer 季夏
Material	Wood	Fire	Metal	Water	Soil
Five Constants 五常	Humaneness 仁	Ritual 禮	Duty 義	Wisdom 知	Faithfulness 信
Five Duties 五事	Bearing 貌	Vision 視	Speech 言	Listening 聽	Thought 思
Seasonal Ordinance 令	Ordinances of Spring 春令	Ordinances of Summer 夏令	Ordinances of Autumn 秋令	Ordinances of Winter 冬令	
Qi	Wood *qi* 木氣	Fire *qi* 火氣	Metal *qi* 金氣	Water *qi* 水氣	Soil [*qi*] 土[氣]

The "Celestial Patterns" draws on the cosmology of the "Wuxing" in its treatment of the planets. It places aberrations in the motion or color of the Dazzling Deluder into a constellation of signs already established in the "Wuxing." Retrograde Mars becomes readable against and amplified by the broader network of signs it is woven into through its association with the ruler's failure to properly engage in the ritual Duty of Vision. Moreover, while Mars had long been associated with summer, the "Celestial Patterns" specifies that the Dazzling Deluder produces meaningful signs when the "Summer Ordinances" are violated.[52] Declaring that the Dazzling Deluder strays from its path due to a fault in the Duty of Vision or a failure to implement the "Summer Ordinances," Ban Zhao and the Ma brothers relocate the meaning of the sign into the Classics.

The "Celestial Patterns" as a Dynastic History

The "Celestial Offices," like the *Shiji* as a whole, covered a period from antiquity until the adulthood of its primary author, Sima Qian. The

"Celestial Patterns" is part of a dynastic history; it covers only the Western Han dynasty. The "Wuxing" and the "Celestial Patterns" tell the story of the rise and fall of that dynasty. The chronicle of celestial signs at the end of the "Celestial Patterns" contains some four dozen entries that tell of the triumph of the dynastic founder Liu Bang, troubling attempts to usurp power early in the dynasty, the glories and eventual excesses of Emperor Wu, and the rising power of the Wang clan following the death of Emperor Yuan 元 (r. 48–33).

The chronicle effectively begins with the Chunqiu period, reproducing in different language events that appear in the "Celestial Offices" through the rise of the Han. The "Celestial Patterns" chronicle then provides a detailed account of celestial signs through the end of the Western Han.[54] A typical entry reads: "In the sixth year of Emperor Wu (135 BCE), the Dazzling Deluder guarded Cart Ghost.[55] The sign-reading said: 'It is an aberration of fire. There will be a funeral.' This year, in the Exalted Garden there was a fire. Dowager Empress Dou died" 六年, 熒惑守輿鬼。占曰:「為火變, 有喪。」是歲高園有火災, 竇太后崩。[56] Four distinct units may be identified in this brief passage: (1) the time of occurrence; (2) the features of the celestial sign; (3) a citation of a sign-reading text; and (4) events that transpired in the terrestrial or human realm.

Toward the end of the Western Han dynasty, increasingly inauspicious signs occurred that, read in retrospect, presage its fall. The final mention of Saturn in the chronicle, in 27 BCE, tells of the Dazzling Deluder following the Quelling Star westward across the sky during the last ten days (xun 旬) of the tenth month. In the first week of the eleventh month, Mars and Jupiter passed Saturn, all of them in retrograde motion, toward the northwest. The sign-reading predicted "warfare and mourning at court and abroad" 外內有兵與喪 and "re-establishment of kings and dukes" 改立王公. Major political upheavals, the chronicle explains, took place over the next two years: the execution of King Xin 歆 of Yelang 夜郎 for treason, the burning of public temples, the theft of imperial seals, and finally, the death of King He 賀 of Liang.[57]

The "Celestial Patterns" records the dramatic appearance of a "flowing star" that glowed red and white and stretched across the sky, more than 100 degrees in length, in the fourth month of 12 BCE. Smaller bright objects, perhaps meteors issuing from the comet's tail, fell behind it. The chronicle interprets this sign both with respect to a similar sign recorded in the *Annals* Classic and with respect to the coming fall of the Western Han:

郡國皆言星隕。《春秋》星隕如雨為王者失勢諸侯起伯之異也。
其後王莽遂顓國柄。王氏之興萌於成帝〔時〕, 是以有星隕之變。

In the commanderies and kingdoms, all said that they were falling stars. In the *Annals*, "stars falling like rain" is a portent of a king losing power and the local lords raising up a hegemon. After this occurred, Wang Mang took control of the empire.[58] The Wang clan began to flourish during the reign of Emperor Cheng, and on account of this, the sign of the falling stars occurred.[59]

The penultimate entry cites the appearance of a bright, long-lasting comet near Oxherd. It seemed to confirm the claims of Xia Heliang 夏賀良 (d. ca. 5 BCE), who cited the age-old axiom that broom-stars swept away the old and ushered in the new in a memorial arguing that the charge of the Western Han needed to be renewed.[60] The court instituted a number of reforms aimed at renewing the charge, establishing a new reign period and imperial title and changing the number of divisions in the clepsydra from 100 to 120. The abolition of Xia's reforms directly presaged the rise of Wang Mang:

八月丁巳, 悉復蠲除之, 賀良及黨與皆伏誅流放。其後卒有王莽篡國之禍。

On the *dingsi* (54) day of the eighth month, all of these reforms were abolished. Xia Heliang and his clique in all cases submitted to punishment and were exiled. Following this, the calamity in which Wang Mang usurped the empire finally occurred.[61]

Xia and his allies must have appeared remarkably prescient in the Eastern Han. The Western Han did fall, and its mandate did indeed need to be renewed. This did not occur, however, until the rise of the Emperor Guangwu 光武 (r. 25–57), a distant relative in the Liu clan, but a Liu nonetheless.

Both the treatise and the dynasty effectively end with the final entry in the chronicle:

元壽元年十一月, 歲星入太微, 逆行干右執法。占曰: 「大臣有憂, 執法者誅, 若有罪。」二年十月戊寅, 高安侯董賢免大司馬位, 歸第自殺。

In the eleventh month of the first year of the Yuanshou reign period (2–1 BCE), the Year Star entered Grand Tenuity, and moved in retrograde toward the vicinity of the Right Enforcer of the Law.[62] The sign-reading said: "The great ministers have cause for worry, and the Enforcers of the Law are punished, as if they themselves were culpable." On the *wuyin* (15) day of the tenth month of the second year, the Marquis of Gao'an, Dong Xian (d. 1 BCE), resigned his post as Marshal of State, returned to his domicile, and committed suicide.[63]

Wang Mang is not mentioned explicitly in the chronicle's final entry, but Dong Xian's suicide nonetheless coincides with the end of Liu power in the Western Han. Dong's fall came with Emperor Ai's death, after which he was accused of refusing to provide the young emperor with medical aid.[64] The next Marshal of State, *de facto* ruler for the remainder of the Western Han, and founder of the next dynasty, the short-lived Xin 新 (9 CE–23 CE), was Wang Mang himself.

Conclusion

The chronicle at the conclusion of the "Celestial Patterns" is emblematic of the shift in the position of celestial signs between the *Shiji* and the *Hanshu*. The "Celestial Patterns" at a glance appears to be a near facsimile of Sima Qian's "Celestial Offices"; a great deal of technical material is shared between the two treatises. However, for Sima, that material is the patrimony of men like himself and his father, readers of celestial signs who had served sages, or been ignored by mediocrities, since high antiquity. Narratives recounting appearances of celestial signs gained their authority through their association with authoritative figures more than their inclusion in authoritative texts. The "Wuxing," and the works of Liu Xiang and Liu Xin on which it was perhaps based, drew not only authority but their very structure from the *Annals* and traditions surrounding the "Great Plan." All manner of signs were organized into categories drawn from the "Great Plan"—the *wuxing*, the Five Duties, and the Sovereign's Standard. Each type of sign presented examples drawn from the *Annals*, in chronological sequence, followed by examples in the Qin and Western Han. While Liu Xiang's "Discourse on the Wuxing Tradition of the 'Great Plan'" could not have included the rise of Wang Mang, Ban Gu's revision in the form of the "Wuxing" did, making each chronicle of each type of sign point toward the ultimate fall of the Western Han. Ban Zhao and the Ma brothers effectively synthesized the technical materials of the "Celestial Offices" with the teleology and classicism of the "Wuxing." In the "Celestial Patterns," celestial signs directly reflected the failure of the ruler to fulfill the Five Duties; technical knowledge became dependent on classical authority, and records of the appearance of celestial signs became a chronicle of the downfall of a dynasty.

Part II
Rhetoric in Practice

Chapter 4

Classical Authority and the Elimination of Baleful Signs

Just as classical authority came to pervade technical treatises from the late Western Han, it likewise became a critical component of memorials to the throne. Memorialists employed authoritative texts to show how baleful signs came into being and how they could be resolved. Their memorials gave accounts of the origins of particular signs, identifying the pathways through which they were produced and pointing toward people, administrative policies, or ritual failures that had produced them. Arguments about how baleful signs and their consequences could be eliminated followed from claims about etiology. Like Master Yan, memorialists claimed that addressing omens meant coming to terms with their origins. Baleful signs and the events they presaged could be resolved—made to melt away—if the underlying conditions that produced them changed. Memorialists appealed to the authority of the Classics, technical texts, and historical exemplars to explain why certain signs came into being and what should be done about them.

Authoritative Theories: From Etiology to Elimination

Memorials on celestial signs routinely cite four of the Five Classics: the *Documents*, the *Changes*, the *Odes*, and the *Annals*. Most citations allude to precedents, historical examples in which a sign—celestial or otherwise—has appeared before. When inauspicious signs occur, the exemplary ruler identifies their etiology and then adjusts his behavior and policies, so that ill-fortune may be eliminated. The inept ruler ignores inauspicious signs and ultimately faces the consequences. In addition to precedents, many memorials cite two classical texts that provide an underlying theoretical basis for the reading of celestial signs: the "Tradition of the Appended Statements" in the *Changes* and the "Great Plan" chapter of the *Documents*.

The "Appended Statements"

"The Appended Statements" provides a cosmological theory that explains why signs come into being. A version of it was included in the Mawangdui tombs, and it had become a fundamental source of classical authority in Eastern Han technical treatises.[1] The "Appended Statements" claimed that Heaven made signs visible but that sages were responsible for reading and interpreting them.[2] Especially after the late Western Han, allusions to and direct citations of the "Appended Statements" became a common feature of rhetoric surrounding celestial signs.

Describing the etiology of baleful signs allowed memorialists to launch attacks on rival factions, political enemies, and close associates of the ruler. Correspondences between the sun and the ruler, and the moon and ministers or palace women, were built on cosmic dynamics central to the "Appended Statements." The political history of the Eastern Han in large part consists in factional struggles between powerful distaff families, a pattern well established by the time of the dynasty's third ruler, Emperor Zhang 章 (r. 76–89). Empresses came primarily from clans of supporters of the founding emperor, most prominent among them Ma Yuan 馬援 (14 BCE–49 CE) and Dou Rong 竇融 (15 BCE–62 CE). Following the death of the Dowager Ma (40–79), Empress Dou 竇 (d. 97) succeeded in promoting the emperor's fourth son to the position of heir-apparent. The new heir was born to a woman of the Liang 梁 clan, longtime supporters of the Dous. The following year, however, a rivalry between the Liangs and Dous emerged, the boy's mother and aunt died (perhaps by suicide), and the Liangs were banished to the southern reaches of the empire.[3] Celestial signs reflected and were read against these complex social and political dynamics.

In 92, the year after he reached the age of majority, Emperor He 和 (r. 88–106) determined to strip the Dou clan of power. In a memorial on an eclipse that summer calling for action against the Dowager Dou and her family, Chancellor Ding Hong 丁鴻 (d. 94 CE), an expert in the Ouyang 歐陽 interpretation of the *Documents*,[4] cited the fundamental symmetry between the sun and moon and the ruler and his ministers. He explained that the sun, like the ruler, should never be diminished, whereas the moon regularly waxes and wanes.[5] When the sun appears to wane, as in a solar eclipse, the fundamental relationship between the ruler and his subordinates is out of balance. Ding explains: "Solar eclipses occur when ministers take power over their lords, and *yin* becomes dominant over *yang*. When the moon becomes full and does not diminish, the lower ranks fill with arrogance" 故日食者, 臣乘君, 陰陵陽; 月滿不虧, 下驕盈也.[6] Ding Hong echoed the language of the "Appended Statements," suggesting that celestial signs were delivered as warnings: "When the ways of men below grow treacherous, verifications appear in the heavens above. Though there may be secret scheming, the spirits illuminate the truth of the matter. The signs are displayed and the warnings made apparent, so as to counsel the lord of men" 人道悖於下, 效驗見於天, 雖有隱謀, 神照其情, 垂象見戒, 以告人君.[7] The *Annals* recorded thirty-six eclipses, Ding Hong ominously noted, and thirty-two assassinations of rulers.[8] Ding Hong claimed that celestial signs revealed otherwise hidden machinations of those serving the emperor, insinuating that the regent Dou Xian 竇憲 (d. 92), the Dowager's eldest brother, had conspired to murder the ruler. Soon after the memorial's delivery, Dou Xian was stripped of his title and committed suicide along with his brothers. The "Appended Statements" suggested that the meaning of a given sign could be understood only by tracing it back to its source. In practice, identifying that source required careful attention to court politics.

The correspondence between sun and ruler, moon and ministers, held true even in arguments for leniency toward imperial officials or members of the imperial household. Li He 李郃 (fl. 89–126 CE), whose expertise included reading trigrams, planets, and wind, had come to prominence after being sponsored by Deng Zhi 鄧騭/陟 (d. 121), brother of the Dowager Deng, as Filial and Incorrupt (*xiao lian* 孝廉). By 120, however, Li had fallen out with the Dengs and been accused of graft, only to be rehabilitated and made Chancellor by Empress Yan 閻 (d. 126) following the death of Emperor An 安 (r. 106–25) in early 125.[9] The fortunes of the Yan clan shifted, however, when the young child they had placed on the throne died in December of the same year.[10] Li argued that the newly enthroned Emperor Shun 順 (r. 125–144) should show leniency toward the dowager, who had ordered the murder of the emperor's mother soon after his birth in 115 and had him deposed from the position of crown prince in 124. In the last year, a number of baleful planetary portents had occurred, suggesting that nobles and officials would be punished and that violence would take place in the women's quarters.[11] Li claimed that celestial signs stood for the speech of Heaven: "I have heard that Heaven does not speak, but displays signs so as to manifest good and ill-fortune, and that it displays disasters, aberrations, and anomalies to reprimand and admonish" 臣聞天不言, 縣象以示吉凶, 挺災變異以為譴誡.[12] Li's memorial appealed to a wide range of past celestial signs, including a bow in the heavens, a guest star, comets, and a record of the moon growing teeth and devouring a star. Drawing on the "Appended Statements," Li drew a close analogy between the appearance of celestial signs and warnings, suggesting that Heaven intentionally produced these signs for the benefit of human beings. He wrote: "Heaven deliberately manifests aberrations so as to clearly demonstrate [its intentions] to people" 天故挺變, 明以示人.[13] Appealing to a precedent in which Duke Huan 桓 of Qi (r. 685–643) had successfully responded to a baleful sign by avoiding the women's quarters, Li suggested that the dowager be kept at a distance from power but treated with respect and provided with due material comforts. Repaying his debt to the Yans without sacrificing his own skin, Li concluded by claiming that those who ignored baleful signs were bound to face calamity, but that sincere regret could forestall misfortune.

While Li's rhetoric treated omens as the product of an intentional Heaven, others cast baleful signs as the result of cosmic imbalances. Zhou Ju 周舉 (d. 149), an expert in the Classics who joined Li He's office in 125 and argued in favor of leniency toward the dowager, opened a memorial on a drought circa 134 with a citation from the "Appended Statements" that positioned the human realm within the broader cosmos. Zhou connected the drought to disturbances in *yin* and *yang* produced by official corruption and excessive expenditures:

臣聞易稱『天尊地卑, 乾坤以定』。二儀交構, 乃生萬物, 萬物之中, 以人為貴。故聖人養之以君, 成之以化, 順四節之宜, 適陰陽之和, 使男女婚娶不過其時。

I have heard it said in the *Changes* that "Heaven is exalted and Earth is humble, and so Qian and Kun are fixed in their places." Through the interaction of the two emblems, the myriad things are born, and among the myriad things, human beings are the most noble. Thus, the sage employs the lord to nourish them and transformation to complete them, causing them to follow along the normative patterns of the Four Seasons and to attune themselves to the harmony of *yin* and *yang*, making it so that men and woman do not miss the opportune time for marriage.[14]

In 132, Liang Na 梁妠 (d. 150), a grandniece of Emperor He's mother, had been named Empress, but after two years had born no heir. In Zhou's memorial, the problem of marriage lies at the beginning of a sequences of causes. Each causal link is greater in magnitude than the next, in a form of rhetoric that might be termed a *chain of amplification*.[15]

Zhou Ju employs this rhetorical form to describe the way blockages in *yin* and *yang* ultimately produce baleful celestial signs, including floods and droughts. What begins as a hidden and incipient problem in the conduct of the emperor and those closest to him manifests on a cosmic scale:

夫陰陽閉隔，則二氣否塞；二氣否塞，則人物不昌；人物不昌，則風雨不時；風雨不時，則水旱成灾。

When *yin* and *yang* are closed off and isolated, then the two types of *qi* are blocked up. When the two types of *qi* are blocked up, human beings and creatures do not flourish. When human beings and creatures do not flourish, wind and rain are unseasonal. When wind and rain are unseasonal, floods and droughts become disasters.[16]

Zhou Ju indicates the ruler and his court[17] as the agents responsible for creating policies aligned with the normative balance of *yin* and *yang*. *Yin* and *yang* are "closed off and isolated," Zhou Ju subtly suggests, because of the many women and men who cannot enjoy marital relations when the harem becomes overcrowded. While the emperor had taken ritual action to alleviate the drought, Zhou argued that concrete changes to policy were more important. Urging the emperor to follow policies of austerity reminiscent of the courts of Emperor Wen and Emperor Guangwu and the precedent of King Wu of Zhou, who released palace women following his conquest of the Shang, Zhou Ju claimed that "by this spirit calamity can truly be transformed into good fortune" 以精誠轉禍為福.[18]

The "Appended Statements" was by no means the only authority cited in the memorials above. Due appreciation is owed to the ways in which it worked in tandem with other types of sources. Crucially, each memorialist studied the current circumstances that faced the court and paid attention to past precedents. Other texts, including both Classics and masterworks, provided historical support to their claims. Yet, the "Appended Statements," above all other texts, served to provide a basis in the Classics for identifying *yin* and *yang* correspondences between the heavens and the human realm, the cosmos and the court. The ruler, configured as male, *yang*, and corresponding to the sun, stood in opposition to his ministers and the women of his court, all of whom were configured as female, *yin*, and corresponding to the moon. Memorialists employed cosmic dynamics to describe not only correspondences but also processes. Irregularities at court became amplified through the cosmos, and minor deviations from the proper patterns of human behavior worsened over time. The "Appended Statements" provided weighty language that bolstered arguments in which memorialists traced baleful signs to their ritual and administrative roots.

The "Great Plan"

Where the "Appended Statements" provided an authoritative source for making claims about cosmic dynamics, the "Great Plan" tied the reading of signs to the duties of kingship. The "Great Plan" is a treatise of divine origin on good government that, according to its preface, could be revealed and transmitted only to the worthy. According to its opening narrative section, King Wu of Zhou, humble and anxious to govern well, sought the advice of worthy men soon after his conquest of Shang. Jizi 箕子 (trad. 11th cent BCE), a former Shang official who had feigned madness to escape the wrath of the wicked King Zhow, offered King Wu a set of instructions for administration that constituted the "Great Plan."

Following its preface, the "Great Plan" details a series of Nine Divisions that comprehensively describe the duties, resources, and methods of governance (see box 4.1). Three of these—the *wuxing*, Five Duties, and the Sovereign Standard—gave structure to the *Hanshu* "Wuxing Treatise." Duties of governance are a central concern of the "Great Plan." These include both the ruler's personal comportment and broader responsibilities. The Five Duties specify how he should conduct himself: looking, listening, and speaking carefully, maintaining a dignified bearing, and thinking deliberately about problems. More broadly,

Box 4.1
The Structure of the "Great Plan"

Preface

The Nine Divisions

The Five Resources/Phases 五行

Water, fire, wood, metal, earth

The Five Duties 五事

Bearing, speech, vision, listening, thought

The Eight Concerns of Governance 八政

Food, money, sacrifices, the duties of the Sikong 司空, Situ 司徒, and Sikou 司寇, guests (diplomacy), the army

The Five Measures 五紀

Year/Jupiter 歲, moon/month, sun/day, stars and planets, astronomical calculations 歷數

The Sovereign Standard 皇極

The Three Virtues 三德

Uprightness 正直, firm control 剛克, soft control 柔克

Resolving Doubts 稽疑

The Many Proofs 庶徵

Rain, sunshine, heat, cold, wind

The Five Blessings 五福 and the Six Extremes 六極

Long life, prosperity, health and peacefulness, cultivation of fine virtue, the ultimate fulfillment of the charge
Misfortune and early death, disease, worry, poverty, vileness, weakness

the *wuxing* specified material resources he was to manage. The Eight Concerns of Governance laid out the broad areas for which the court was responsible. Three of these, the Sikong, Situ, and Sikou, loosely corresponded to the titles given to the highest-level ministers in the late Western Han; in 8 BCE, the titles Da Sikong 大司空, Da Situ 大司徒, and Da Sima 大司馬 were applied to the roles of Chancellor, Imperial Counsellor, and the Grand Marshal.[19] The "Five Measures" specified a set of astronomical objects to observe, all closely tied to keeping good time. The "Great Plan" concerned not only the duties of governance but also its proper methods. The text provided means of controlling subjects ranging from soft power to force in the "Three Virtues," a set of goals to be achieved in the "Five Blessings," and a set of ills to be avoided in the "Six Extremes." It provided a set of weather-related signs to watch for in the "Many Proofs," and a method for selecting a course of action in the face of conflicting information—in the form of advice, intuition, and sign-readings—in "Resolving Doubts."

Memorialists made arguments concerning how signs came into being and how they could be eliminated, often appealing to the "Great Plan" as a standard to which governance should return. Memorialists knew multiple Classics and had the skill to use them together to address complex problems. Etiology and elimination went hand in hand.

The "Great Plan" served to identify and resolve cosmic imbalances. In 26 CE, the year following Emperor Guangwu's accession, Yin Min 尹敏 (fl. ca. 26–ca. 72), of Guangwu's home commandery of Nanyang 南陽, argued that the new emperor and his court should use the "Great Plan" to understand and resolve baleful signs. An expert in the Ouyang interpretation of the *Documents* and both the *Guliang* 穀梁 *Tradition* and *Zuo Tradition* of the *Annals*, Yin pointed to a process in which the Six Extremes of the "Great Plan," including ill-fortune, disease, and so forth, could be traced back to a series of Six Disruptions (*liu li* 六沴), disturbances in *qi* associated with the Five Phases in the "Wuxing Treatise":

六沴作見，若是供御，帝用不差，神則大喜，五福乃降，用章于下。若不供御，六罰既侵，六極其下。明供御則天報之福，不供御則禍災至。欲尊六事之體，則貌、言、視、聽、思、心之用，合六事之揆以致乎太平，而消除轥軻孽害也。

When the Six Disruptions occur, if the proper offerings are made, the ancestors will not sigh,[20] and the spirits will be overjoyed. The Five Blessings will thereupon be sent down, illuminating all below. But if the proper offerings are not made, the Six Penalties will advance, and the Six Extremes will descend. Whosoever understands how to make proper offerings, Heaven will reward with blessings. But calamity and disaster will come to those who do not make proper offerings. If one desires to fulfill the Six Duties, then in enacting bearing, speech, vision, listening, thought, and mind (xin 心), he must bring those Six Duties into a single standard of measurement so as to bring about Great Peace, resolving and eliminating the difficulties of baneful injury.[21]

The *Hanshu* "Wuxing Treatise" uses the term *li* 沴 to describe one type of *qi* exercising a harmful, destabilizing effect on another. Wood and metal disrupt one another, as do water and fire, and in some instances, metal, wood, water, and fire all disrupt the central phase, soil.[22] The "Wuxing" explained "When [different types of] *qi* do harm to one another, this is called *li*" 氣相傷, 謂之沴.[23] The Six Disruptions manifest signs that demand appropriate responses. Both successful and unsuccessful responses lead directly into the last of the Nine Divisions of the "Great Plan," either the Five Blessings or the Six Extremes. Success hinges upon fulfilling Six Duties. These are the same as the Five Duties of the "Great Plan" with thoughtful mind (*sixin* 思心) split into two, "thought" (*si*) and the "mind" (*xin*). At a time when civil wars raged on and Emperor Guangwu had yet to gain the support of Ma Yuan or Dou Rong, Yin argued that fulfilling the Six Duties would bring peace and eliminate baleful signs. Avoiding the Six Extremes and receiving the Five Blessings required the ruler to identify their source in the Six Duties, and thereby rectify ritual faults of his own or of those close to him. Yin Min's memorial earned him a place as a Gentleman of the Palace (*langzhong* 郎中) in Emperor Guangwu's court.[24]

Other early Eastern Han writers likewise appealed to the "Great Plan" as a means of eliminating imbalanced cosmic conditions responsible for the appearance of baleful signs. General Zhu Fu 朱浮 (d. 57 CE), an early supporter of Emperor Guangwu, employed etiological arguments based on *yinyang* dynamics and cited the "Great Plan" as a means of identifying and responding to the conditions that had produced a solar eclipse in 30. Power dynamics among Guangwu's followers had shifted dramatically in only a few years, as Ma Yuan and Dou Rong pledged their support to the emperor in 28 and 29.[25] Officials in the commanderies were frequently dismissed over seemingly minor lapses, perhaps owing to concerns about their abilities or their loyalty to the new regime.[26] Citing a *Documents* precedent in which the sage-king Shun subjected matters of promotion and demotion to "three-fold examination" (*san kao* 三考), Zhu argued that this state of affairs made it so that officials lost their posts before they could learn to effectively fulfill their duties, thereby producing unease and uncertainty among local populations.[27] Zhu pointed to the parity between the ruler and the sun:

臣聞日者眾陽之所宗, 君上之位也 . . . . . . 若陽上不明, 尊長不足, 則干動三光, 垂示王者。

I have heard that the sun is the most honored of the many [kinds of] *yang* and corresponds to the superior position of the lord. . . . If *yang* above is not bright, but lacking in honor and esteem, then the Three Luminaries [i.e., the sun, moon, and stars] will be disturbed and send down their admonitions to the king.[28]

Where *yinyang* dynamics explained the nature of the problem, following models in the Classics offered a solution. Zhu continued: "The Five Classics record the policies of states and great clans. The 'Great Plan' distinguishes the patterns in disasters and anomalies. All proclaim and illuminate the Way of Heaven, so as to verify the affairs that are to come" 五典紀國家之政, 鴻範別災異之文, 皆宣明天道, 以徵來事者也.[29] Gesturing toward incipient events that the signs might presage, Zhu referenced the "Great Plan" as a whole rather than a particular passage within it. Used in conjunction with records elsewhere in the Classics, the text worked as a comprehensive guide for understanding the relationship between policy and the appearance of celestial signs. Because these undefined events were not set in stone, the capacity of the court to respond to the signs left the future open. The "Great Plan" served to identify patterns in celestial phenomena and to explain how ill-fortune could be eliminated.

A memorial from the following year makes a similar case, likewise using both *yinyang* dynamics and the "Great Plan" to argue that an early solar eclipse indicated that Emperor Guangwu needed to improve his management of officials. Zheng Xing 鄭興 (1st cent CE), a *Zuozhuan* expert who had once studied under Liu Xin, used the occasion to argue that the emperor should appoint Guo Ji 郭伋 (38 BCE–47 CE) as Imperial Counsellor. Guo enjoyed wide support among court officials, and like Zheng Xing was an associate of Ma Yuan. Zheng's memorial described an interwoven cosmos, in which disasters in the heavens and monstrosities on the earth all result from human actions. In general terms, Zheng explained, "When human beings turn against virtue there is disorder, and when there is disorder, monstrosities and disasters are born" 人反德為亂, 亂則妖災生.[30] Zheng claimed that disasters in the sun and the moon occur due to fundamental failures in the basic three duties of governance, the selection of officials (*ze ren* 擇人), providing a basis for the people (*yin min* 因民), and following the seasons (*cong shi* 從時).[31] Arguing that Emperor Guangwu should heed his ministers' recommendations in making the highest-level appointments, rather than appointing his sister's husband, Li Tong 李通 (d. 42),[32] Zheng urged him to follow the precedent of the sage-king Yao and other exemplary rulers in "following the hearts of the people" (*yin ren zhi xin* 因人之心) when making decisions about major political appointments.[33] Pointing to trepidation among officials at Emperor Guangwu's court, Zheng characterized the eclipse as a warning from a benevolent Heaven that sought to guide the ruler lest he stray from his proper path:

夫日月交會, 數應在朔, 而頃年日食, 每多在晦。先時而合, 皆月行疾也。日君象而月臣象, 君亢急則臣下促迫, 故行疾也......今陛下高明而羣臣惶促, 宜留思柔剋之政, 垂意洪範之法, 博採廣謀, 納羣下之策。

As for encounters between the sun and the moon, they regularly occur on the first day of the lunar month, but eclipses of the sun in recent years have in many cases occurred on the last day of the month. When they come together before the proper time, it is in every case due to the hurried movement of the moon. The sun is the counterpart of the lord, and the moon, the counterpart of his ministers. When the lord is high-handed and insistent, his ministers below are in straitened circumstances, and so, their movements are pressed. . . . Now, Your Majesty is exalted and perspicacious, but your many ministers are wracked with fear. It is fitting that you consider a policy of soft control, concentrating upon the methods of the "Great Plan." Graciously choose among their far-reaching plans, and accept the proposals of your many subordinates.³⁴

Zheng Xing inverted the typical understanding of ministers usurping the power of the sovereign in his reading of the eclipse, thereby encouraging the sovereign to listen to and yield to his ministers. He appeals to Emperor Guangwu to exercise "soft control" (*rou ke* 柔克), one of Three Virtues in the "Great Plan," rather than force to bring his court to order. While *yinyang* dynamics might be understood as the mechanism that amplified human relations into a cosmic event, the "Great Plan" allowed Zheng to tie failures in administration to standards that laid out, in specific terms, what good governance meant.

Memorialists used *yinyang* dynamics and the "Great Plan" to explain the appearance of a wide variety of signs, constructing elaborate chains of amplification to identify their sources and indicate how they might be eliminated. Following the death of the Dowager Deng in 121, Emperor An took personal control of the reins of the administration for the first time. After an earthquake and period of "excessive rain" (*chang yu* 常雨) the following year, Chen Zhong 陳忠 (d. 125) wrote a memorial protesting the dismissal of Chen Bao 陳褒 (fl. 102–122), the former Imperial Counsellor (Sikong 司空) who had been appointed in 120, when the dowager had still been in power. Chen Zhong cited the "Five Duties of the 'Great Plan'" (Hongfan wushi 洪範五事) by name, claiming that excessive rain resulted from a failure to maintain reverence and solemnity in the Duty of Bearing (*mao* 貌). He claimed that in the *Annals* great floods occurred when rulers lacked majesty. Ministers serving under such rulers grew arrogant, and *yin qi* burgeoned. Poor governance disturbed *yinyang* dynamics, ultimately leading to baleful signs:

臣聞位非其人，則庶事不敘；庶事不敘，則政有得失；政有得失，則感動陰陽，妖變為應。

I have heard that when the wrong people occupy [high-ranking] positions, various [administrative] affairs go out of order. And when the various affairs are out of order, in governance there are errors. When in governance there are errors, this disturbs *yin* and *yang*. Monstrosities and aberrations are the response.³⁵

While Chen Zhong evoked the notion of resonance, *ganying* 感應 (stimulus and response), by describing monstrosities and aberrations as *ying* 應 (responses) to poor government, he employed a chain of amplification to create the sense that problems in the human realm exist prior to those responses. Chen argued that the emperor should restore power to the Three Excellencies and that the disasters resulted from the emperor allowing close connections, including his in-laws, his former wet-nurse, and her daughter, to gain too much power: "The handles (of power) that belong to Your Majesty are in the hands of your ministers and wives. The occurrence of floods and disasters surely began with this" 陛下之柄在於臣妾。水災之發，必起於此。[36] The emperor's former wet-nurse and her daughter were exiled to the northern frontier, but Chen Bao never regained his position.[37]

The court of Emperor An's only son, Emperor Shun, would come to be dominated by the family of his empress, the Liang clan. The situation persisted well into the reign of Emperor Huan 桓 (r. 146–168), until 159, when a group of eunuchs aided the emperor in ousting the Liangs from power. Tensions grew between the eunuchs and other factions at court in the ensuing years, ultimately coming to a head in the crisis of 168. In 166, Xiang Kai, an astute reader of both texts and omens who held no official position, had presented a pair of memorials defending two local officials, Liu Zhi 劉質/瓆 (d. 166) and Cheng Jin 成瑨 (d. 166), who had been imprisoned after executing eunuchs and their supporters in Nanyang and Taiyuan 太原. Xiang Kai cited a litany of signs, passages from various Classics, and apocryphal texts associated with the *Classic of Filial Piety*, the *Changes*, and the "Great Plan."[38]

Rafe de Crespigny, in his classic study and translation of Xiang Kai's biography, cites no fewer than twelve types of signs across the two memorials, including planetary anomalies, extreme weather events, the collapse of gates at the imperial academy, eclipses, and the emperor's failure to produce a son (perhaps resulting from the baleful influence of the emasculated eunuchs).[39] Trained in Jing Fang's interpretation of the *Changes*, Xiang cites a "Tradition of the *Changes*" (*Yi zhuan* 易傳) throughout his memorials, as well as *Changes*-related texts generally identified as apocrypha, the *Yi tian ren ying* 易天人應 (Responses between Heaven and the Human World for the *Changes*) and the *Yi zhongfu zhuan* 易中孚傳 ("Inner Trust" Tradition for the *Changes*).[40] Xiang likewise cites apocrypha associated with the *Classic of Filial Piety* and the "Great Plan" in his discussion of a near conjunction between Jupiter and Venus in the lodges House and Heart that had occurred the previous year:

房、心者,天帝明堂布政之宮。孝經鉤命決曰:「歲星守心年穀豐。」尚書洪範記曰:「月行中道,移節應期,德厚受福,重華留之。」重華者,謂歲星在心也。今太白從之,交合明堂,金木相賊,而反同合,此以陰陵陽,臣下專權之異也。

House and Heart correspond to the Celestial Emperor's Hall of Illumination and the palace from which he carries out his administration. The *Xiaojing gou ming jue* (Recipe for Obtaining [lit. hooking] the Charge for the *Classic of Filial Piety*) says: "When the Year Star guards Heart, the year's grain harvest will be plentiful." The *Shangshu Hongfan ji* (Notes/Records on the "Great Plan" of the *Venerable Documents*) says: "When the moon moves along the central path, its regulated movements resonate with measures of time, so that Virtue is abundant, blessings are received, and Zhonghua retains it." As for Zhonghua, this refers to the Year Star being in Heart. Now, Great White has followed it, and there has been a conjunction between them in the Hall of Illumination. Metal and Wood do violence to one another, yet return to join in unison. This is because *yin* overcomes *yang*; it is an anomaly corresponding to ministers below who dominate power at court.[41]

Xiang's use of the *Shangshu Hongfan ji* raises questions concerning the relationship between the apocryphal text and the *Documents* itself. These concern both content and authority. In what ways did the apocryphal texts develop, depend upon, or even enhance the rhetorical power of the Classics? Working with the passage at hand, the answers to these questions must remain tentative and incomplete. In terms of content, it must be noted that the "Great Plan" includes both the moon (or month) and the Year (or Year Star, Jupiter) among its Five Measures, as well as the cultivation of virtue among its Five Blessings. The "Great Plan" does not precisely indicate, however, how the month or the year, the moon or Jupiter, is to be observed, nor does it mention Zhonghua, an alternate name for the sage-king Shun. The apocryphal citation suggests that the text developed the "Great Plan" by tying it to more specific methods of observation and interpretation, while also introducing new correspondences. The *Xiaojing gou ming jue*, by contrast, has no clear relationship to the content of the *Xiaojing* itself. In both cases, the apocryphal texts have a complex, interdependent relationship with the Classics. Xiang's use of the *Shangshu Hongfan ji* invests the "Great Plan" with newer, more specific, and more esoteric technical knowledge that issued from a mysterious source. This in turn allows Xiang Kai to underscore his cosmological claims concerning the excessive power of the eunuchs at Emperor Huan's court with the hallowed authority of the Classics.

The "Great Plan" entered into the reading of all manner of baleful signs, celestial and otherwise, as factional tensions continued under Emperor Ling 靈 (r. 168–189), the last Han sovereign to reign before the empire collapsed into civil war. Following Emperor Huan's death in 168, his last empress, Dou Miao 竇妙 (d. 172), appointed her father Dou Wu 竇武 (d. 168) as General-in-Chief (Da jiangjun 大將軍). Supported by the Grand Tutor (Tai fu 太傅) Chen Fan 陳蕃 (d. 168), Dou Wu attempted to move decisively against the eunuchs, hoping to execute their leaders. The action failed, Dou Wu committed suicide, and his supporters were killed or forced into exile. Those that survived, dubbed the Proscribed Faction (*dang gu* 黨錮), were banned from government service.[42] In 178, Emperor Ling, by then a young man in his early twenties, had deposed his empress and executed her father and brothers, leaving their bodies to rot unburied. After a series of baleful signs that included fog, comets, an earthquake, an epidemic, and an eclipse, the Palace Attendant Lu Zhi 盧植 (d. 192) cited the "Great Plan" as the key to responding to such signs. A former student of Ma Rong responsible for producing commentaries on the *Documents* and the *Rites*, Lu paired an understanding of the etiology of ill portents of all types with the means to eliminating them. Lu's etiology uses both *wuxing* and *yinyang* dynamics, characterizing power at court in terms of fire and water, *yin* and *yang*.

臣聞漢以火德, 化當寬明。近色信讒, 忌之甚者, 如火畏水故也。
案今年之變, 皆陽失陰侵, 消禦災凶, 宜有其道。

I have heard that the Han employs the virtue of fire, and therefore its transformative influence ought to be broad and bright. Yet keeping close to attractive favorites and trusting in slanderers is the worst violation of ritual standards, for fire recoils from water. In examining the aberrations of recent years, we find that all are instances of *yang* losing its position and *yin* encroaching. Yet there is a means by which disasters can be eliminated and ill-fortune prevented.[43]

Lu argued that many members of the Proscribed Faction banned from governance for the last decade were innocent and should be pardoned. He claimed that the unburied corpses of kinsmen of the deposed empress had caused recent epidemics and that excessive taxes had led to banditry. Lu urged the emperor to "cultivate the rites" (*xiu li* 修禮) by summoning worthy officials to court, such as the disciples of Zheng Xuan. These worthies would be able "to enact and illuminate the 'Great Plan' and to exorcise and pacify disasters and signs of blame" 陳明洪範, 攘服災咎.[44] Emperor Ling—the last Eastern Han sovereign to be more than a pawn under the power of a warlord—was unmoved by his words.

In many of the memorials we have seen, *yinyang* dynamics and the "Great Plan" are used in conjunction with one another. The "Appended Statements" is often, but by no means always, cited in discussions of *yinyang* dynamics, and no doubt silently lent authority to this type of cosmological argument, even where it was not explicitly mentioned. What differences do we see in the ways in which the "Great Plan" and the "Appended Statements" are deployed in memorials? What distinct types of claims do these texts authorize?

The "Appended Statements" provides an authoritative source for making claims regarding *yinyang* dynamics. The ruler is invariably tied to *yang*, while his ministers, his attendants, his in-laws, his empress, and any empress dowager are tied to *yin*. Signs that involve excesses of *yin* or *yang*, such as floods, droughts, eclipses, and earthquakes readily map onto imbalanced power dynamics at court. Ding Hong's memorial against Dowager Dou after an eclipse, Zhou Ju's argument that keeping too many women in the harem had led to drought, and Cheng Zhong's claim that excessive rains signaled that Emperor An's in-laws and the family of his nursemaid had gained too much power all work from the fundamental resonance between *yinyang* dynamics at court and in the broader cosmos.

Whereas the "Appended Statements" (and *yinyang* dynamics in general) worked to map cosmic disturbances onto power imbalances, the "Great Plan" authorized claims regarding more specific failures in the duties of kingship. Its Nine Divisions laid out a schematic vision of the various aspects of good governance, covering topics that included resource management, timekeeping, comportment at court and during rituals, sign-reading, and decision-making. The Eight Concerns of Governance (*ba zheng*) allowed memorialists to point to specific failures with regard to finances, sacrifices, military matters, and even the duties of the Three Excellencies. The Five Duties could be used—as had been done systematically in the "Wuxing Treatise"—to point to particular types of ritual failures at court, to say that the ruler or members of his court had failed to deliberate on the problems of governance, to listen carefully to voices of loyal officials, or to maintain a bearing appropriate to high office. The "Great Plan" and the notion of the *wuxing* (in Eastern Han, variably meaning Five Resources or Five Phases) also allowed for discussing the etiology of cosmic disturbances in more complex terms than binary *yinyang* dynamics allowed. The *wuxing* readily mapped onto seasons (five in total, including late summer) and the Five Planets, among other things. *Wuxing* dynamics were well suited to linking together large numbers of disparate signs, such as planetary omens, comets, strange clouds, and portents involving diseases or aberrant animals. However, in relating these signs back to power at court, *wuxing* dynamics were sometimes translated into *yinyang* dynamics. Lu Zhi paired water and fire with *yin* and *yang*. Faced with the planetary omen of Great White (Venus) following the Year Star (Jupiter) into Heart, Xiang Kai moved from describing Metal encroaching on Wood to *yin* overcoming *yang* and eunuchs dominating the ruler at court.

The "Appended Statements" and the "Great Plan" were authoritative sources that memorialists could draw upon to make cosmological claims regarding the etiology of all manner of signs, auspicious and inauspicious, celestial and otherwise. These two texts presented malleable theories that allowed memorialists to contextualize baleful signs with respect to both the conditions that produced them and the events they presaged. Memorialists connected signs to one another, to conditions at court, and to conditions in the empire as a whole, producing powerful arguments that linked developments in the political and religious center of the human world to cosmological processes. As useful a rhetorical tool as theory was, however, it was by no means the only one memorialists had at their disposal. Memorialists appealed also to precedents in which baleful signs had appeared in the past and to the models of exemplary rulers who had addressed those signs by rectifying themselves or their courts.

Authoritative Exemplars: Turning Calamities into Good Fortune

Explaining how baleful signs came into being was a powerful suasive technique but only insofar as something could be done about them. Indeed, for a baleful sign to function as a call to action, the action must at least have the potential to be efficacious. Explanations of etiology allowed the memorialist to point toward the conditions that brought baleful signs into being. If those conditions were eliminated, so too would be the sign itself and whatever catastrophic events it might have presaged.

Claims that baleful signs were subject to change rested on authoritative precedents. Memorialists looked to occurrences of baleful signs under sage rulers in the distant past to show how the correct response to such events could turn calamities into good fortune. Certain stories were told and retold to the point where simply mentioning the names of the persons involved evoked whole narrative arcs. Judging from extant sources, three precedents proved particularly valuable in making arguments concerning celestial signs: King Wuding of Shang's self-rectification after a pheasant appeared on the ear of his tripod; King Cheng of Zhou's acknowledgment of the hidden virtue of the Duke of Zhou 周公 (11th cent. BCE) following an omen in which wind blew over trees; and Duke Jing of Song's selfless refusal to sacrifice his chief minister, his subjects, or the bounty of the harvest to exorcise an ill portent.

What was it about these stories that made them so resonant with both memorialists and rulers? How did their rhetoric achieve its effects? In a general sense, these narratives gained their power by pointing to exemplary rulers who responded properly to the appearance of baleful signs and reaped the rewards. Yet questions remain. Were they cited for specific details or images, or for the broader meanings that emerged from their overarching narratives? How did memorialists evoke them—did they name specific chapters, draw upon their language as proof-texts, or refer to the memorable images they contained? Most crucially, we may understand something of what the narratives achieved by considering them through counterfactuals: If a given precedent did not exist, would it have been possible to write essentially the same memorial?

King Wuding and the Pheasant in "The High Ancestor's Day of Sacrifice"

"The High Ancestor's Day of Sacrifice" chapter of the *Documents* presents King Wuding of Shang's exemplary response to an inauspicious omen. Chapter 2 examined the role this narrative plays in the "Basic Annals" of the *Shiji*.[45] The *Documents* text reads:

高宗肜日，越有雊雉。祖己曰：「惟先格王，正厥事。」乃訓于王，曰：「惟天監下民，典厥義。降年有永有不永。非天夭民，民中絕命。民有不若德、不聽罪,天既孚命正厥德,乃曰其如台。嗚呼!
王司敬民。罔非天胤,典祀無豐于昵。」

On the High Ancestor's Day of Sacrifice, a pheasant cried on the ear of the tripod. Zuji said: "It has come before the king to rectify his affairs." And he admonished the king, saying: "Heaven watches over the king's men below, supporting them in their duties. And while the years we are allotted are in some cases many and in others few, it is not Heaven that causes men to die young, cutting their lives (*ming*) short while they are still in their prime. There are many among the people who do not yield to your virtue, who refuse to acknowledge their crimes. Heaven bestows its charge (*ming*) on those who rectify their own virtue! Yet now they say, 'What are we to do?' Alas, as you sit upon the throne, you must be reverent toward your subjects. None among the ancestors was not a scion of Heaven.[46] Do not give too abundantly to your father[47] in your sacrifices."[48]

In the "Basic Annals," the reign of King Wuding was a renaissance for the Shang dynasty, just as the reign of King Xuan was for the Zhou. Zuji advised the king not to make excessive sacrifices to his father, or perhaps his nearest relatives, lest he fail to properly honor the more ancient Shang kings. The *Documents* chapter itself does not explain that King Wuding heeded Zuji's sage advice, but memorials describe King Wuding's reign as inaugurating a century of prosperity,[49] and he fittingly received the auspicious temple name "High Ancestor" (Gaozong 高宗).

In 19 BCE, Marshal of State Wang Yin 王因 (d. 15 BCE), Emperor Cheng's maternal uncle and perhaps the most powerful official in the empire, presented a memorial on a flock of pheasants that descended during the Grand Archery Ceremony (*Da she li* 大射禮). After more than a decade on the throne, Emperor Cheng had yet to produce an heir, and worse still, had begun leaving the palace in disguise, often accompanied by his male favorite, Zhang Fang 張放 (d. 7 BCE), to enjoy the pleasures of nightlife in Chang'an. Wang pointed to the "Day of Sacrifice" as a precedent showing that an inauspicious sign could be used as an opportunity for reform and renewal:

天地之氣，以類相應，譴告人君，甚微而著。雉者聽察，先聞雷聲，故月令以紀氣。經載高宗雊雉之異，以明轉禍為福之驗。

The *qi* of Heaven and Earth mutually respond among objects of like-kind, so as to reproach and warn the lord of men, in a subtle yet clear fashion. If we examine the pheasant by listening, first we hear the sound of thunder, for the monthly ordinances serve to give order to *qi*. The Classic records the anomaly of the pheasant that cried at the court of the High Ancestor of Shang, so as to clearly illuminate the verification of how ill-fortune can be turned into good.[50]

Wang Yin combined the appearance of the gathered pheasants with the sound of thunder to integrate both into the dynamic interaction of *qi* between Heaven and Earth. The image of the birds appearing during the ceremony evoked the precedent in a way that no other sign could. However, soon after the appearance of the sign, the palace attendant Chao Hong 鼂閎 (fl. 19 BCE) called its authenticity into question, noting that damage to the birds' wings indicated recent captivity. Nonetheless undeterred, Wang warned the emperor against listening to Chao's concerns, characterizing his speech in the most inauspicious terms as "words that might fell a nation" 亡國之語.[51] Moreover, he explained the specific etiology of the sign, offering Emperor Cheng a method by which misfortune might be eliminated:

外有微行之害，內有疾病之憂，皇天數見災異，欲人變更，終已不改......宜謀於賢知，克己復禮，以求天意，繼嗣可立，災變尚可銷也。

Outside the court, you have subjected yourself to dangers in your incognito travels. At court, you are beset by worries of sickness and disease. August Heaven many times has manifested disasters and anomalies, in the hope that you might turn over a new leaf, but up to this point, you've yet to change your ways. . . . You ought to heed the counsels of the worthy and the knowledgeable, take control of yourself, and return to ritual propriety, so as to seek Heaven's Will! Your heir can be established; disasters and aberrations still can be resolved.[52]

Wang Yin's conclusion underscores the notion that a proper response to the signs would eliminate both the signs themselves and the circumstances that had produced them. Whether or not the appearance of the pheasants was spontaneous—and therefore genuine—the circumstances that had produced the sign were real enough. Emperor Cheng, like King Wuding of Shang, might yet renew his dynasty should he reform himself.

Kong Guang 孔光 (d. 5 CE), writing under Emperor Ai, likewise cited the "Day of Sacrifice" to argue that the proper response to an inauspicious sign could lead to blessings and renewal. Kong had been made Chancellor at Emperor Cheng's death, but lost his position in 5 BCE due to his outspoken claims that the Dowager Empress Fu 傅, the emperor's grandmother, and other members of her clan exercised undue influence over the ruler.[53] Responding to a 2 BCE eclipse and the subsequent death of the dowager some two weeks later,[54] Kong Guang described the etiology of the eclipse using the conventional analogy between the ruler and the sun in which a solar eclipse signals the encroachment of *yin* forces on the ruler, so that "the virtue of the lord is diminished" 君德衰微.[55] Drawing on the language of the "Great Plan," Kong Guang treated the eclipse as a sign that the ruler was in precarious position:

如貌、言、視、聽、思失，大中之道不立，則咎徵薦臻，六極屢降。皇之不極, 是為大中不立，其傳曰『時則有日月亂行』。

When bearing, speech, vision, listening, and thought are all lost, and the Supreme and Central Way is not established, then the proofs of blame all arrive and the Six Extremes descend one after another. When the Sovereign fails to find the Standard (*huang zhi bu ji*), this is a failure to establish the Supreme and Central Way. The tradition [attached to the "Great Plan"] says "When this occurs, the sun or moon moves chaotically."[56]

It is the "Day of Sacrifice," however, that Kong Guang deployed to underscore the notion that the sign was a warning calling on the ruler to reform:

書曰『惟先假王正厥事』，言異變之來，起事有不正也。臣聞師曰，天（右）〔左〕與王者，故災異數見，以譴告之，欲其改更。

The *Documents* says, "It came before [the events it presaged] to tell the kings to set right their affairs,"⁵⁷ meaning that when anomalies and aberrations come about, there is something that is not right that brings them into being (*qi shi*). I have heard from my teacher that Heaven assists the king, and therefore many times manifests disasters and anomalies to reproach him, desiring that he mend his ways.⁵⁸

Kong Guang treats Heaven as a benevolent agent that produces signs for the benefit of the ruler. The wise ruler seeks to understand their etiology, "that which brings them into being." Once their etiology is understood, the signs can then be resolved. Citing factionalism, slander, and avarice at court, Kong Guang argued that "worthy and good"⁵⁹ (*xian liang* 賢良) officials must be employed, punishments and fines moderated, and taxes reduced. Near the conclusion of his memorial, Kong Guang again appealed to the "Day of Sacrifice": "The *Documents* says: 'Heaven bestows its charge on those who align their virtue!' This speaks of aligning your virtue so that you might follow the course of Heaven"⁶⁰ 書曰『天既付命正厥德』, 言正德以順天也. Kong Guang emphasized that rites of exorcism would be a useless response to the eclipse.⁶¹ Only by addressing the conditions that produced the sign could the emperor "resolve the calamity and make good fortune flourish" 銷禍興福,⁶² just as the High Ancestor once had done.

Kong Guang and Wang Yin used the precedent in two rather different ways. In Wang Yin's case, the image of the pheasants appearing during the sacrifice was critical, and it was the image—rather than the specific language of the *Documents*—that Wang invoked in his memorial. Kong Guang, by contrast, twice cited specific language from the text. The narrative itself was sufficiently well known that Kong mentioned neither the name of the chapter nor any of the figures involved. Operating in the background may have been a subtle hint that the emperor should not dwell too much on the recent death of his grandmother, just as Zuji warned that Gaozong should not offer excessive sacrifices to his father. If this is the case, however, it was never explicitly stated. The critical aspect of the precedent lies instead in the general truths it asserts: the best of rulers work to rectify themselves, and Heaven responds in kind.

King Cheng and the Violent Winds

The "Metal-bound Coffer," in contrast to the "Day of Sacrifice," places the problem of the slandered official front and center. The year after the

Classical Authority and the Elimination of Baleful Signs | 127

Zhou conquest, King Wu grew ill, and his younger brother, the Duke of Zhou, pleaded with his ancestors, King Tai 大王, Wang Ji 王季, and King Wen, to take his own life in exchange for allowing his older brother to live. Having received an auspicious response, the Duke of Zhou sealed away the strips (*ce* 冊) recording his communications with the ancestors in a metal-bound coffer, and the king recovered. The narrative advances from this moment to the chaotic times following the king's death a few years later. Around that time, the Duke of Zhou's surviving brothers, led by Guan Shuxian 管叔鮮 (d. ca. 1041 BCE)[63] rebelled, spreading rumors that the Duke of Zhou would soon depose the king and place himself on the throne. The Duke of Zhou then spent two years in the east, quelling their rebellion. Upon the Duke of Zhou's return from his campaign against his rebellious brothers, King Cheng did not punish him, but regarded him with suspicion and fear:[64]

> 王亦未敢誚公。秋，大熟，未穫，天大雷電。以風禾盡偃，大木斯拔，邦人大恐，王與大夫盡弁，以啟金縢之書，乃得周公所自以為功、代武王之說。

> The king did not dare censure the Duke. In autumn, there was a bumper crop, but before the grain could be reaped, the heavens shook with thunder and lightning. Wind blew down the stalks of grain and pulled the great trees from the earth. The people of the land were very afraid. The king and the noblemen donned their caps and opened the metal-bound coffer to view the writings. Only then did they learn of the Duke of Zhou's merit, and of his attempt to offer himself as a substitute for King Wu.[65]

Only at this point did King Cheng come to recognize the Duke of Zhou's true character. The king realized also that the violent, unseasonable weather manifested Heaven's recognition of the Duke, in contrast to his own failure to recognize his uncle:

> 王執書以泣，曰：「其勿穆卜。昔公勤勞王家，惟予沖人弗及知。今天動威以彰周公之德，惟朕小子其新逆，我國家禮亦宜之。」

> The king held the writings and wept, saying: "Do not divine about it. In the past the Duke labored for the royal house, but because I was but a little child, I did not understand it.

Now, Heaven quakes and inspires awe in order to illuminate the character (*de*) of the Duke of Zhou. I, the young son, must go and greet him anew. The rites of my domain and my house require it."[66]

The tearful young king's change of heart provoked a heavenly response in turn. "The king went out into the suburbs, and the heavens sent down rain and a wind that blew in the opposite direction, so that the stalks all rose up. The Two Dukes ordered the people of the land to replant all the trees that were lying down. That year, there was indeed a bumper crop" 王出郊, 天乃雨, 反風, 禾則盡起。二公命邦人, 凡大木所偃, 盡起而築之。歲則大熟.[67] The winds that blew in reverse mirrored the reversal in King Cheng's treatment of the Duke of Zhou. The bounty of the harvest was an auspicious sign that signaled Heaven's approval just as the violent winds had signaled Heaven's admonition.

In several instances, memorialists cited violent winds that ripped trees from the ground during the reign of King Cheng of Zhou, sometimes in the same breath as they recalled the pheasant that appeared at King Wuding's court. In a memorial on the 12 BCE appearance of Halley's Comet (discussed in detail in the following chapter), Liu Xiang cited the two precedents in close proximity, using both to underscore the potential for rulers to rectify themselves and their courts. Du Ye 杜鄴 (d. 2 BCE) brought the two precedents to bear on the same 2 BCE solar eclipse that Kong Guang had used to criticize the excessive power of the Fu clan. Du took aim at the clans of both the emperor's grandmother and his mother, the late Dowager Ding 丁 (d. 5 BCE). Using a combination of precedents and *Changes*-based *yinyang* dynamics, Du argued that women, however prominent or powerful, must remain in the *yin* position and under the control of men. Even the virtuous mother of King Wen had, after all, remained under the charge of her son. By contrast, early in the Western Han, Empress Lü had gained too much power after the founding emperor's death, dominating his heir and filling positions of power with members of her own clan. The gendered nature of Du's rhetoric notwithstanding, his attacks were aimed more at clans than individuals. The emperor's mother had died several years earlier, but members of her clan, like the Fus, continued to accumulate both wealth and rank. Despite the emperor's frugality and attention to ritual, Du notes that he had received no auspicious signs, only solar eclipses and earthquakes, both of which issued from topsy-turvy *yinyang* dynamics at court:

> 日食，明陽為陰所臨，坤卦乘離，明夷之象也。坤以法地，為土為母，以安靜為德。震，不陰之效也。

> Solar eclipses show that *yang* is being imposed upon by *yin*. When the Kun ☷ trigram presses upon the Li ☲ trigram, this is the image of Mingyi 明夷 (Diminishing Brightness ䷣). Kun ought to emulate the earth, acting as soil and as mother, taking repose and rest as its virtue. Quakes (*zhen*) are the verification that it is not acting in a manner befitting *yin*.[68]

How did selecting this particular hexagram support Du's broader arguments? Kun, plainly enough, maps onto *yin* and allows Du to associate an idealized notion of feminine virtue with the stillness and repose of the earth. Li, associated with fire, maps onto the ideal of embodying the virtue of fire. The interaction of the two trigrams is also at play. Earthquakes occur when the soil behaves in an aberrant fashion, trembling when it ought to be still. The Mingyi hexagram, moreover, contains within it the trigram *zhen* ☳, meaning "quaking" or "thunder." Du uses the *Changes* well, creating a sense that prevailing interactions of *yin* and *yang*, Kun and Li, and soil and fire are responsible for social and cosmic disturbances. Near the conclusion of Du's memorial, he claims that baleful signs are an aid and warning to the ideal ruler, citing both the precedents of King Wuding and King Cheng to remind the emperor that bad fortune can indeed be turned to good:

> 天變不空，保右世主。如此之至，奈何不應！臣聞野雞著怪，高宗深動；大風暴過，成王怛然。願陛下加致精誠。

> Heaven does not produce aberrant signs in vain, but does so to protect and aid the lord of the age. When such signs have come about, how can you fail to respond? I have heard that the High Ancestor of Shang was deeply moved by the appearance of the omen of the wild bird, and that King Cheng of Zhou grew terrified after the violent passage of the great winds. Would that Your Majesty become so perfect in integrity.[69]

As Du Ye employed them, a single lesson was to be learned from the two precedents. Howsoever baleful the signs a ruler confronted, it was his own response to them that mattered most.

Though the "Metal-bound Coffer" itself never mentions the death of the Duke of Zhou, Han readings sometimes understood King Cheng's recognition of the Duke of Zhou to be posthumous.[70] In the year following Dou Wu and Chen Fan's ill-fated attempt to eliminate the eunuch faction at court, a number of strange portents occurred, including the appearance of a blue snake before the throne, hail, lightning, great winds, and trees being ripped from the ground. In the summer of 169, Zhang Huan 張奐 (104–181) employed these as a basis for arguing that leniency should be shown to the members of the Proscribed Faction who had been banned from politics and that those who died in the violence should be afforded proper burial:

> 昔周公既薨，成王葬不具禮，天乃大風，偃木折樹。成王發書感悟，備禮改葬，天乃立反風，其木樹盡起。今宜改葬蕃、武，選其家屬，諸被禁錮，一宜蠲除，則災變可消，昇平可致也。

> Long ago when the Duke of Zhou died, King Cheng did not follow the full rituals in burying him. Heaven thus sent forth a great wind that knocked down trees and bent their trunks. King Cheng then found what the Duke of Zhou had written, had a sudden realization, and reburied him in accord with the full rituals. Heaven then sent down another wind that blew in the opposite direction, so that the trees stood up once more. Now, we ought to rebury Chen Fan and Dou Wu with full honors and nominate their heirs. And as for all those who have been proscribed from politics, they should be pardoned at one stroke. Then the disasters and aberrations can be resolved, and peace can be had.[71]

Zhang Huan himself had commanded the Northern Brigade against Dou Wu on behalf of the eunuch leader Cao Jie 曹節 (d. 181 CE) in the previous year, a decision he seems to have regretted,[72] perhaps having discovered some hidden virtue of Dou Wu and Chen Fan after their deaths, just as King Cheng had discovered the Duke of Zhou's act of self-sacrifice when he opened the metal-bound coffer. Zhang Huan presented a powerful case for giving Dou Wu and Chen Fan a proper burial based on the authoritative classical precedent of King Cheng's posthumous honors for the Duke of Zhou. The emperor, still a boy of thirteen *sui* who had yet to begin personal rule, is said to have been deeply moved by Zhang's memorial but powerless to enact his recommendations.

As was true of King Wuding and the pheasant, marked contrasts appear in the ways that different memorialists deployed the precedent of King Cheng and the violent winds. Du Ye (and, as will be seen in chapter 5, Liu Xiang) used the latter precedent to assert the same general truth as the former. Du (and Liu) cited the two precedents together to show that the circumstances behind recent baleful signs might indeed be undone with due effort. Du responded to an eclipse that bore no special relationship to the signs in either *Documents* chapter. In Zhang Huan's memorial, the particular details of the "Metal-Bound Coffer" were, however, much more salient. Zhang cited strong winds knocking down trees in his own day, just as had occurred following the death of the Duke of Zhou. Still more critical, there are close parallels in the circumstances that brought the signs into being. Dou Wu, like the Duke of Zhou, was a former regent, regarded as a potential usurper, and treated dishonorably after his death. However, due to the entrenched power of the eunuch faction at court, Emperor Ling, unlike King Cheng, was unable to ritually honor the dead.

The Integrity of Duke Jing of Song

While King Cheng responded to a series of baleful celestial signs by honoring the Duke of Zhou's offer to sacrifice himself in exchange for King Wu, Duke Jing of Song refused to sacrifice his Prime Minister, his people, or the bounty of the harvest to save himself from the baleful influence of the Dazzling Deluder. The well-known story, perhaps taking place in 494, 480, 479, or 472 BCE, is not found in any of the Five Classics but was nonetheless part of the common repertoire of anecdotes known to those versed in classical learning.[73] The *Shiji* recalls:

熒惑守心。心，宋之分野也。景公憂之。司星子韋曰：「可移於相。」景公曰：「相，吾之股肱。」曰：「可移於民。」景公曰：「君者待民。」曰：「可移於歲。」景公曰：「歲饑民困，吾誰為君！」子韋曰：「天高聽卑。君有君人之言三，熒惑宜有動。」於是候之，果徙三度。

> The Dazzling Deluder [i.e., Mars] guarded the Heart constellation. Heart is the celestial field corresponding to Song. Duke Jing grew afraid because of it. Zi Wei, the Observer of Planets, said: "Its [baleful influence] can be transferred to the Prime Minister." Duke Jing said: "The Prime Minister is my top aide." Zi Wei said, "It can be transferred to the people." Duke Jing said, "A lord relies on his people." Zi Wei said, "It can be transferred to the harvest." Duke Jing said, "If the harvest is poor, the people will be in dire straits. Over whom should I then be a lord?" Zi Wei said, "Heaven is exalted (*gao*, lit. high) but listens to those who are humble (*bei*, lit. low or lowly). You, My Lord, have three times spoken the words of a lord. The Dazzling Deluder ought to move." Thereupon, they observed it. As expected, it moved three degrees.[74]

Other well-known versions of the narrative conclude in still more dramatic fashion. In the *Huainanzi*, Duke Jing explained that he would rather sacrifice his own life than his kingdom: "If I wanted to kill my people in order to survive, who would consider me a true lord? My life (*ming*) has certainly reached its end. Zi Wei, speak no more" 而欲殺其民以自活也, 其誰以我為君者乎? 是寡人之命固已盡矣, 子韋無復言矣.[75] The *Lüshi chunqiu* 呂氏春秋 (Annals of Lü Buwei, comp. ca. 239 BCE) version uses nearly the exact same language.[76] Immediately after hearing this moving speech, Zi Wei turned to the north and told the Duke that he had effectively earned a pardon. In both the *Huainanzi* and the *Lüshi chunqiu*, the life of the ruler is likewise saved by his refusal to make self-serving sacrifices, but the Dazzling Deluder moves three full lodges, rather than the more plausible three degrees of the *Shiji*.[77] While the passage does not spell out the relationship between the ruler and the celestial sign, Heaven clearly responds to his words and intentions.

A version of this narrative took place in the late Western Han, this time with a tragic twist. Its record in the "Celestial Patterns" treatise may serve as a basic outline:

> 二年春, 熒惑守心。二月乙丑, 丞相翟方進欲塞災異, 自殺。
>
> In spring of the second year (Jan. 31–Apr. 29), the Dazzling Deluder guarded Heart. On the *yichou* (2) day of the second

month (Mar. 27), Chancellor Zhai Fangjin (d. 7 BCE), desiring to put a stop to the anomaly, committed suicide.[78]

Some weeks later, the emperor died as well.

Zhai Fangjin's *Hanshu* biography tells a more complex version of the story. The sign appeared, and Li Xun 李尋 (d. after 5 BCE), who worked under Zhai, submitted a memorandum (*ji* 記) to the throne that included a number of signs. In addition to "Mars guarding a lodge" 火守舍, Li pointed to aberrant signs involving Venus and Saturn, appearances of comets, signs involving mountains and rivers, and rumors that had spread among the populace. He laid blame on both the emperor and his officials, warning that high-ranking ministers were in a treacherous position:

上無惻怛濟世之功，下無推讓避賢之效，欲當大位，為具臣以全身，難矣！

He who is above has not the merit of showing mercy for the world and rescuing it, while those below show no evidence of yielding to make way for worthies, desiring instead to maintain their own positions of power, act as prominent ministers, and thereby keep their bodies whole. Yet this proves difficult indeed![79]

Zhai's biography notes that this worried him, and he did not know how to deal with the situation. One Fei Li 賁麗 (fl. 7 BCE), an expert in planetary omens seemingly aware of sacrificial means of answering such signs similar to those that Zi Wei had cited, suggested that "a great minister ought to block it" 大臣宜當之.[80]

Emperor Cheng then issued a statement that denounced Zhai on a number of charges, noting the frequent appearance of disasters since he had taken office, as well as famine, epidemics, and civil strife among the populace in recent years. The emperor blamed Zhai for high taxes and policies leading to insufficiency even when harvests were bountiful, concluding with a scathing reprimand:

欲退君位，尚未忍。君其孰念詳計，塞絕姦原，憂國如家，務便百姓以輔朕。

I have desired to demote you, but up to this point haven't had the heart to do so. Would that you make a careful accounting, stop up the source of this treachery, worry about the empire as much as you do your own household, and take as your task aiding me in succoring the hundred families.[81]

Emperor Cheng then gave Zhai an ox and a great quantity of wine, gifts that might be used to conduct a sacrifice. That very day, Zhai committed suicide. The emperor provided him with a lavish funeral, overseeing the ceremony in person, and the posthumous title "Honored Marquis" (*gong hou* 恭侯).[82] If the point of the original narrative was to suggest that the ruler could save himself and his domain through an act of self-sacrifice, Emperor Cheng's sacrifice of Zhai Fangjin and his own subsequent death strengthened that precedent by way of an inverse example.

One question that arises in conjunction with this event is whether it may have been fabricated. In an influential study of the events surrounding the death of Zhai Fangjin, Huang Yi-long noted that Mars did not move through Heart until August, well after the springtime deaths of both Emperor Cheng and Zhai Fangjin, and did so in prograde fashion. Mars did, however, move retrograde back through Horn (*jiao* 角, in Virginis) from early February through mid-April, returning to prograde motion a few days after Emperor Cheng's death on April 17.[83] We know that Mars was not in Heart in spring of 7 BCE, yet both Zhai Fangjin's biography and the "Celestial Patterns" treatise say that it was. Is it possible that someone at court claimed that Mars was in retrograde motion in Heart and that no one discovered this was not the case? This seems highly doubtful. Identifying the current location of a planet in the heavens requires only the most rudimentary knowledge of the constellations, and its location would have been obvious to many of those present at Emperor Cheng's court. Identifying the past location of a planet was much more difficult. Recall that Li Xun said only that "Mars guarded a lodge," but did not name a particular lodge. The reference to Heart comes in narration surrounding the memorial, supplied by the biography's compiler(s), rather than in language attributed to figures present at Emperor Cheng's court. We may speculate that, at some point, Mars' true location slipped away from the historical record and the narrative force of the precedent of Duke Jing of Song served to tell later historians where Mars *must have been*.

Chen Zhong's memorial following the earthquake and heavy rains of 122, discussed above for its use of the "Great Plan," employed both the original precedent and a historical argument concerning Emperor Cheng's failure to follow it. Concerned with the increasing power of the Imperial Secretariat (Shang shu 尚書) as opposed to the Three Excellencies since Dowager Deng's death, Chen Zhong cited Duke Jing's refusal to sacrifice his own Prime Minister:

近以地震策免司空陳褒，今者災異，復欲切讓三公。昔孝成皇帝以妖星守心，移咎丞相，使賁麗納說方進，方進自引，卒不蒙上天之福，徒乖宋景之誠。故知是非之分，較然有歸矣。又尚書決事，多違故典，罪法無例。

On account of a recent earthquake, the Imperial Counsellor Chen Bao was dismissed from his post. Now, on account of disasters and anomalies, you are also about to dismiss the Three Excellencies in one fell swoop. Long ago, the Filial Emperor Cheng, facing the appearance of a monstrous star guarding Heart, transferred the blame to his Chancellor, having Fei Li place the blame on Zhai Fangjin, who then committed suicide. In the end, Emperor Cheng did not receive the blessings of High Heaven, for he had simply turned his back on the integrity of Duke Jing. Knowledge of the right and wrong decision lies along a clear path. Moreover, the Secretariat has in many cases violated hallowed precedents in handling affairs, failing to distinguish between culpability and lawfulness.[84]

Chen Zhong used the occasion not only to defend Chen Bao but also to criticize the Secretariat. His memorial presents a political twist on the magical rites of exorcism Zi Wei offered. Zi Wei proposed to transfer the baleful influence of the Dazzling Deluder, without considering what had caused the sign or who was to blame.[85] In Chen's memorial, however, the dangers of the sign are inseparable from the conditions that caused it. Chen claimed that Emperor An had ignored the true cause and wrongly blamed Chen Bao. In allowing an innocent minister to suffer, the emperor had made the very mistake Duke Jing might have made, but did not. Chen Zhong issued a subtle yet grave warning to Emperor An, pairing

Duke Jing's exemplary actions with Emperor Cheng's poor response to the similar omen in the late Western Han. When the Dazzling Deluder guarded Heart[86] and Emperor Cheng blamed his Chancellor, he had brought misfortune on himself and died without progeny soon thereafter.[87]

The story of Duke Jing's exemplary response to the Dazzling Deluder bolstered arguments that rulers could recover good fortune from baleful signs. Several years before penning his memorial on the circa 134 drought, Zhou Ju produced a memorial concerning an encounter between Mars and the moon, in which he defended Chancellor Zhu Chang 朱長 (d. after 127) against charges that old age had made him incapable of carrying out his duties.[88] Zhou's rhetoric emphasized the ruler's capacity to forestall ill-fortune by overcoming the conditions that produced the sign. In Zhou's view, Heaven sent the sign to aid the ruler:

楚莊曰：『災異不見，寡人其亡。』今變異屢臻，此天以佑助漢室，覺悟國家也......宋景公有善言，熒惑徙舍，延年益壽......務知戒慎，以退未萌。

King Zhuang of Chu (r. 613–591) once said, "Disasters and anomalies no longer appear—I am finished!" Now, aberrations and anomalies arrive with great frequency. These are Heaven's aid to the House of Han, sent to awaken the ruling house.... Duke Jing of Song said his fine words, and the Dazzling Deluder moved to another lodge, so that his years were extended and his life was lengthened.... Take wisdom as your task, be cautious and circumspect, and so reverse [ill-fortune] before the sprouts have grown.[89]

Zhou Ju employed the exemplar of Duke Jing in arguing for the corrective potential of the sign. By raising the case of Duke Jing in conjunction with King Zhuang of Chu's famous declaration that when signs ceased to appear all hope for the sitting ruler was lost, moreover, Zhou Ju highlighted the salutary effect of baleful signs. Baleful signs carried a doubled valence. They announced both that the ruler was in a dangerous position and that he had the capacity to extricate himself from it.

A memorial by Cai Yong 蔡邕 (132–192) on a series of planetary omens, likely dating to 182 and perhaps heavily abridged,[90] pointed to dangerous conditions signaled by the movement of Mars in Grand Tenuity and the visibility of Venus during the day. These signaled strong states becoming weak, weak states becoming strong, and the absence of effective governance. The situation demanded careful attention to ritual, military matters, and administration. The exemplar of Duke Jing suggested that the ruler should begin by cultivating his own virtue:

> 臣聞熒惑示變，人主當精明其德。則有休慶之色......以杜漸防萌，則其救也。昔宋景公小國諸侯，三有德言，而熒惑為之退舍。

> I have heard that when the Dazzling Deluder manifests aberrations, if the ruler of men purifies and illuminates his virtue, then fine and auspicious colors will prevail.... If you stop up the gradual and guard against the sprouts, then we will be saved. Long ago Duke Jing of Song, local lord of a small domain, three times spoke words of virtue, and so, the Dazzling Deluder moved retrograde through the lodges.[91]

The memorial claimed that planetary omens originate in the excessive power of ministers. Employing the exemplar of Duke Jing, it focused attention on the potential for corrective action rather than on the origins of the sign. Any remaining possibilities for recovering the dynasty's fortunes would be lost, however, by the end of the decade. Emperor Ling's death in May 189 marked the end of the Eastern Han in all but name.

Among these memorials, which aspects of the precedent of Duke Jing's response to the Dazzling Deluder proved to be most salient? Was the precedent cited for its specific imagery, its structural features, or perhaps its language? Those at Emperor Cheng's court seemed not to be concerned with this particular narrative, but with the more immediate problem of what Mars in retrograde motion might mean at a time when the emperor was already in poor health and had recently nominated a nephew as his successor, apparently having given up altogether on passing his throne to a son. As Li Xun noted, it was one sign among a great many. Fei Li seemed to allude not to the precedent itself but to practices similar to those Zi Wei had initially recommended, magical means of displacing the malevolent influence of the planet from the ruler onto a minister. More than a century later, Chen Zhong's memorial in defense of Chen Bao denounced unjust treatment of high-ranking ministers. In his response to an earthquake and torrential rains, the particular details of the sign proved less important. Zhou Ju's defense of Zhu Chang similarly cautioned against unjust treatment of a worthy minister, though it responded to a sign that bears some formal similarity to the precedent, an event in which Mars and the moon came into close proximity. Cai Yong cited the sign in response to other planetary omens, appealing only to its most general lesson: one can recover good fortune by properly responding to baleful signs. It is worth noting that none of these memorials cited the sign for its language alone; none drew upon it as a proof-text. The precedent was not, after all, recorded in the Classics, however memorable and rhetorically powerful classicists found it to be.

Conclusion

The "Appended Statements" allowed memorialists to claim that baleful signs were cosmic reflections of imbalances in *yinyang* dynamics in the human world, while the "Great Plan" provided a framework for making claims about standards of good governance and the duties of kingship. What, then, did precedents achieve that the "Appended Statements" and the "Great Plan" could not? They lent the gravity of historical memory to claims that there was something paradoxically auspicious in each baleful omen, for disasters and anomalies represented Heaven's aid to the ruler, so that he might rectify his own ritual comportment, his court, or the administration of his empire. The precedent of King Wuding's response to the appearance of the pheasant on the cauldron lent powerful support to arguments in favor of personal and administrative reform, including those criticizing excessive favoritism toward close relatives of the emperor. King Cheng's ultimate recognition of the Duke of Zhou bolstered arguments that high officials who had fallen from favor be honored once again, whether this meant a proper burial or a return to high office. The narrative surrounding Duke Jing of Song's response to the appearance of the Dazzling Deluder in the Heart constellation speaks, in particular, to the notion that the baleful influence of a given sign was not fixed but transferable and subject to change. The narrative suggests there were ritual methods of removing the planet's baleful influence and shifting it onto persons or things in the realm other than the ruler himself, yet it also cautions against relying on ritual alone. The auspicious response comes not from employing such methods, but from the ruler taking personal responsibility for the sign in his refusal to redirect its influence. Duke Jing of Song's response to the Dazzling Deluder served as a solemn reminder that the best of rulers did not shift blame onto their subordinates but placed the welfare of the realm ahead of even their own lives.

Chapter 5

Reading Signs, Reading the World

Reading celestial signs meant identifying the relationship between signs and conditions in the human world. The great classicist Yang Xiong described two kinds of sign-reading, one in which readers observed Heaven to understand the human world, and another in which readers observed the human world to understand Heaven:

> 或問聖人占天乎？曰占天地。若此，則史也何異？曰史以天占人，聖人以人占天。

> Someone asked me: "Do the sages read signs in the Heavens?"
> "They read signs in Heaven and Earth."
> "If that is so, how do they differ from court archivists?"
> "Court archivists use Heaven to read human affairs. The sages use human affairs to read Heaven."[1]

Reading Heaven was inseparable from an understanding of the human world. Sages—those most deeply attuned to human affairs—anticipated celestial signs before they appeared. When aberrant developments occurred below, corresponding irregularities would manifest above. They corrected the course of things, realigning the human realm with regular celestial patterns even before inauspicious omens could appear. As a fourth-century commentary put it, "the great sages anticipated Heaven, and so, Heaven did not contravene them" 大聖先天而天不違. Court archivists, by contrast, kept records and worked backwards to understand celestial signs in terms of developments in the human realm. Less astute and perspicacious than the sages, they relied on meticulous review of court records, general historical knowledge, and mastery of classical texts to uncover the underlying meaning of baleful signs. Operating in the day-to-day environment of the court, they read Heaven to expose patterns already prevalent in the human world.[2]

Both David Schaberg and Li Wai-yee have examined the rhetorical use of signs in speeches in the *Zuozhuan*, arriving at insights that may be extended to Han memorials. While the heavens are "not moral but regular,"[3] their regularity is the source for human morality, a model the ruler must emulate. Discussions of baleful signs center on the human world. Schaberg writes:

> The world's intelligibility, its whole availability as a source of infallible signs for the rhetorical treatment of human action, is informed by the assumption that it works for virtue and that it is readable on the basis of training in traditional knowledge . . . Although these realms of knowledge are made in some cases to provide a quasi-scientific, extra-human basis for what is right in the human world, they regularly return to the practical considerations of human behavior in society.[4]

Schaberg notes that not only signs themselves but the world as a whole is intelligible, for conditions within the human realm are always read in conjunction with the signs. Thus the reading of signs never becomes a fully systematized, mechanical activity. "[T]he various representations of a single theory (astronomy, for instance) in separate speeches do not necessarily add up to a coherent body of knowledge. System is held in abeyance for the sake of local rhetorical force."[5] Li Wai-yee likewise points to the practice of reading the human realm in conjunction with baleful signs that issue from non-human sources. "Anomalies become the occasion for remonstrances," she writes, "when they are regarded as the manifestations of disorders in the human realm."[6] An anomaly is not a meaningful sign, in other words, unless it can be tied to human failure.[7]

The preceding chapter focused on sources of authority, including commonly deployed theory and precedents, that bolstered arguments centered on etiology and elimination in a broad set of memorials. This chapter examines memorials in three historical moments, each of which shows that the reading of celestial signs was inseparable from the reading of human affairs. The first set concerns a lack of auspicious signs rather than the presence of inauspicious signs. A close analysis of Dong Zhongshu and Gongsun Hong's memorials early in the reign of Emperor Wu suggests that the dearth of memorials on disasters and anomalies in the mid-Western Han may be explained not only by the frequency of auspicious omens during Emperor Wu's reign, as Martin Kern has suggested,[8] but also by rhetorical practices in which memorials critical of imperial policies or ritual protocols addressed a lack of good omens rather than the presence of bad ones. The second set features memorials by Gu Yong and Liu Xiang in the years following the dramatic appearance of Halley's Comet in 12 BCE. Each writer employed classical texts, technical knowledge, and historical precedents yet indicated different paths by which the sign had come into being and different means of eliminating it. The third set includes memorials by Zhang Heng, Ma Rong, and Li Gu 李固 (ca. 94–147) following the earthquake of 133 CE. Zhang pointed to the parity between signs in the heavens and signs in the earth, arguing that recent appearances of both types of signs issued from policy changes concerning which men could be recommended as Filial and Incorrupt. Ma interpreted the earthquake as the product of a failed agricultural policy. Li identified the empire as an extension of the body of the emperor, arguing that seemingly minor mistakes at court could have cosmological consequences. Though the writers of these three sets of memorials worked in different historical circumstances, each read signs as reflections of the world in which they appeared.

Seeking Auspicious Signs at the Court of Emperor Wu

Memorials on celestial signs in the late Western and Eastern Han dynasties predominantly concern inauspicious signs, disasters and anomalies (*zai yi* 災異) rather than auspicious emblems (*rui* 瑞) that proclaimed the legitimacy of the emperor.[9] From the reign of Emperor Yuan through the late Eastern Han such warnings were common, but before that period they were rare. The memorials of Dong Zhongshu and Gongsun Hong pertaining to celestial signs early in the reign of Emperor Wu represent a rhetorical approach to celestial signs distinct from that of late Western Han and Eastern Han memorials. Both writers used celestial signs to make arguments regarding administrative and ritual policy at court. However, unlike those of their later counterparts, their arguments hinged on a lack of emblems of legitimacy rather than the appearance of baleful celestial signs. Nonetheless, Dong and Gongsun, like later writers, employed etiological arguments, claiming that undertaking the actions they suggested would produce the desired results. Both writers addressed the absence of auspicious signs because the edicts that invited their composition asked about the absence of such signs.

Dong Zhongshu responded to a series of three edicts, dating to early in the reign of Emperor Wu. The first was most likely produced in 134 BCE, soon after the death of the Dowager Dou 竇 (d. 135 BCE). Each subsequent edict takes into account Dong Zhongshu's responses. Both the edicts and responses are included in Dong Zhongshu's biography, forming a sort of conversation between Dong and the young emperor. While scholarship over the last several decades has overturned the once-common view of Dong as the progenitor of a "Han Confucian synthesis,"[10] these memorials are the writings that may be most reliably attributed to him. In the initial edict, Emperor Wu posed a series of questions concerning proper governance to perhaps one hundred officers who had been nominated as "worthy and good" (*xianliang* 賢良).[11] This edict expressed concern with the origins of omens bad and good, yet it focused on the question of how to encourage auspicious signs and prosperous realities, asking what the ruler and his court must do to summon a litany of positive omens: sweet dew from the heavens, bountiful harvests, cold and heat in balance, and the Three Luminaries remaining whole (*san guang quan* 三光全).[12] Good signs implied the absence of bad ones; if the Three Luminaries remained whole, then no eclipses occurred, and if cold and heat were balanced, no summer frosts appeared. The positive signs of the ruler's legitimacy followed the receipt of sacrifices to the proper spirits: seasonable weather and agricultural productivity.

The edict that prompted Gongsun Hong's memorial, dating perhaps to the fifth year of Emperor Wu's adult rule (130 BCE),[13] addressed experts knowledgeable in "records of celestial patterns, earthly contours, and human affairs" 天文地理人事之紀,[14] and inquired about the etiology of both auspicious and inauspicious signs. The edict recalled a period in high antiquity (*shang gu* 上古) when, despite a rustic and unceremonious mode of governance, ubiquitous auspicious signs signaled the perfected state of the world and its ruler:

陰陽和, 五穀登, 六畜蕃, 甘露降, 風雨時, 嘉禾興, 朱中生, 山不童, 澤不涸; 麟鳳在郊藪, 龜龍游於沼, 河洛出圖書。

> Yin and yang were in harmony,
> the Five Grains full,
> the Six Domesticated Animals multiplied,
> sweet dew fell,
> winds and rains were seasonable,
> fine grains flourished,
> vermilion grasses grew,
> the mountains did not grow barren,
> the marshes did not dry up,
> the unicorn and the phoenix were in the suburban moors,
> the turtle and the dragon swam in the pools,
> and the Yellow River and the Luo River produced the Chart and the Writings.[15]

Emperor Wu, naturally enough, asked how he might bring such things about in his own time: "I consider these to be fine things indeed. Now, by what path might I bring them here" 朕甚嘉之，今何道而臻乎此?[16] Second, he asked about the etiology of baleful signs that occurred under good rulers, including Yu's work to end the floods under Yao and drought in the time of the Shang founder Tang.[17] Finally, he asked how such signs could have come into being under ideal rulers whose reigns ought to have been visited by auspicious signs alone: "What brings about the abandonment or efflorescence of the Tally of Heaven's Charge" 天命之符，廢興何如?[18] Dong and Gongsun's responses addressed questions about causes. The edicts assumed that the signs appeared due to conditions in the world and placed the ruler at the center of that etiological process.

Dong Zhongshu

Dong Zhongshu's *Hanshu* biography largely comprises his responses to the three imperial edicts dating to circa 134 BCE.[19] Some doubt exists as to the precise extent that the language of the responses is Dong's own as the memorials contain at least two anachronistic place names. Ban Gu may have altered the responses to suit his own time and inclinations, presenting what he believed to be Dong's views[20] but perhaps making rhetorical adjustments. As they have been transmitted, Dong's responses seem to anticipate late Western Han and Eastern Han rhetorical practices, pointing to the *Annals*, the *Odes*, and the *Documents* as ethical guides for the ruler. He argued that the imperial use of harshness and favor must follow the course of the seasonal cycles of yin and yang, and he encouraged restraint in the use of punishment. Finally, he argued for increased use of favor and moral suasion, citing the need to create educational institutions and establish regulations to prevent officials from engaging in production and trade. Dong pointed to signs that accompanied the appearance of great rulers in the past, judging that the lack of the appearance of such signs in the Han evinced that the emperor should reform administrative and ritual policy.

Dong Zhongshu uses the Classics in multiple ways, drawing on them for theories of kingship, as sources of precedent, and as authoritative proof-texts. In a manner reminiscent of the text of the *Gongyang Tradition*, Dong finds deep significance in the seemingly mundane language of the *Annals*. Its first lines carry great weight: "First (*yuan*) year, Spring, the King's first (*zheng*) month" 元年春, 王正月.[21] In Dong's view, these few words describe the ideal relationship between the ruler and Heaven. Dong writes:

臣謹案春秋之文, 求王道之端, 得之於正。正次王, 王次春。春者, 天之所為也; 正者, 王之所為也。其意曰, 上承天之所為, 而下以正其所為, 正王道之端云爾。

I have carefully examined the language of the *Annals*. [Those who] seek the trailhead for the Kingly Way find it in rectification (*zheng*). "Rectification" follows "King," and "King" follows "Spring." Spring is the work of Heaven, and rectification is the work of the King. What it means is this: His Majesty (lit. he above) carries on Heaven's works, while his subjects (lit. those below), owing to this, rectify their own works. This is why rectification is said to be the trailhead for the Kingly Way.[22]

For Dong Zhongshu, the *Annals* Classic is the ultimate authority behind his more detailed and complex cosmological claims regarding *yinyang* dynamics and their relationship to human actions. The *Odes*, by contrast, are primarily used as proof-texts in Dong's rhetoric, with variable attention to the historical circumstances in which individual odes were thought to have been composed. In one instance, Dong cites the *Odes* to reinforce a general claim about good governance:

善治則災害日去，福祿日來。詩云：「宜民宜人，受祿于天。」為政而宜於民者，固當受祿于天。

When good governance is put into practice, disasters and harm grow more distant each day, and blessings and rewards come closer each day. The *Odes* says, "He brings peace to his officers and his people / And receives his reward from Heaven."[23] Those who carry out their administrations in a manner that suits their subjects surely ought to receive rewards from Heaven.[24]

Dong's language mirrors the words of the ode, adding force to his general claim. While early commentary connects it with the court of King Cheng of Zhou, Dong draws on the poem for the general truth it contains rather than to evoke a particular sage figure from the past. Elsewhere, however, Dong cites the *Odes* with greater attention to historical circumstances, including praise for King Wen's circumspect attention to ritual and good governance, and lines lamenting the decline of the Western Zhou.[25]

The *Documents* Classic serves, most critically, as a source of precedent. Recalling the white fish that leaped into King Wu's boat prior to his conquest of Shang, Dong argued that certain types of signs must appear to show the legitimacy of the ruler:

臣聞天之所大奉使之王者，必有非人力所能致而自至者，此受命之符也。天下之人同心歸之，若歸父母，故天瑞應誠而至。《書》曰「白魚入於王舟，有火復於王屋，流為烏」，此蓋受命之符也。周公曰「復哉復哉」，孔子曰「德不孤，必有鄰」，皆積善累德之效也。

I have heard that when Heaven's great emissaries come to a king, they bear those things that cannot come about by human power alone, yet come about all on their own. These are the Tallies of Receiving the Charge. The people of the realm, united in their hearts, take refuge in him, as if they were taking refuge in their own fathers and mothers, and so, Heaven's Auspicious Emblem responds to his integrity and arrives. The *Documents* says: "The white fish entered the Royal Vessel, and the fire in the Royal Chamber transformed into a crow."[26] These must have been the Tallies of Receiving the

Charge.[27] The Duke of Zhou said, "It has come! It has come!" and Kongzi said that "Virtue never dwells alone, but always has neighbors."[28] These are events that verify the piling up of goodness and the accumulation of virtue.[29]

In treating these past signs as "events that verify the piling up of goodness," Dong Zhongshu indicated the role of the ruler in their production. Heaven produced signs that no human power could produce yet issued those signs in response to the actions of the sovereign. King Wu received the Charge, the command to overthrow the Shang, because his own ancestral line had accumulated merit. But that Charge was revealed as such only because it was accompanied by Heaven's Auspicious Emblem (*tian rui*).

The Way, which for Dong referred to the proper method of governance, was ever present and always available to those rulers who would follow it. The decline of a ruling house was not the result of a capricious Heaven; rather, it signaled the failure of men: "When the Way of Zhou fell into decline under King You and King Li, it was not that the Way was lost, but that King You and King Li failed to follow it" 夫周道衰於幽厲, 非道亡也, 幽厲不繇也.[30] The Way was not something that came into being and faded away for no reason. Like a footpath through the undergrowth, it persisted so long as it was traveled.

Dong employed familiar etiological arguments, with human agency remaining at their source. In the late Western Zhou, the failure of King Li and King You to follow the way led to strife between the local lords and enmity throughout the human realm. In times of decline, Dong explained, rulers became licentious and frivolous, relying on harshness rather than favor. This in turn produced malevolent *qi* that disturbed both social harmony and the balance of *yin* and *yang*, leading to disasters and anomalies. Cosmic disruptions issued from human failures. The fall of dynasties and the appearance of baleful signs were both the responsibility of the ruler.

Dong wrote in response to both a lack of signs and a lack of institutions that he considered to be necessary for sagely governance. Emperor Wu's edict sought a general explanation of signs, both auspicious and inauspicious, in the past. In a manner that mirrored his description of the production of inauspicious signs, Dong employed chains of amplification to describe how general prosperity, good governance, and auspicious signs might be brought into being. The ruler had to right his own heart to right his court, right his court to right his subjects, and right his subjects to right the world. In doing so, he could avoid the production of malevolent *qi* responsible for disasters and anomalies, instead bringing about the ultimate signs of cosmic legitimacy:

是以陰陽調而風雨時，羣生和而萬民殖，五穀孰而屮木茂。

This is the means by which *yin* and *yang* will be properly tuned, the winds and rains made seasonable, the many living things brought into harmony, the Myriad Subjects allowed to lay down roots, the Five Grains ripened, and the grasses and trees made to flourish.[31]

Dong claimed that the emperor could win the fealty of far-away nations, bring order to the cosmos, and achieve the Kingly Way by rectifying himself. His noted political opponent and rival *Annals* expert Gongsun Hong[32] likewise responded to the absence of auspicious signs with arguments that located human agency as the etiological source of good omens.

Gongsun Hong

Gongsun Hong's memorial showed that he shared Dong's view on the cosmic resonances of governance and advocated limiting the role of punishment and increasing the role of moral suasion. It is otherwise largely devoid of concrete suggestions for changes in imperial policy; the vague nature of the memorial might have contributed both to the Superintendent of Ceremonial's (Feng chang 奉常) decision to dismiss it as low in quality and to Emperor Wu's decision to declare it the best he had received.[33] The memorial earned Gongsun a private audience with the emperor, during which he may have made policy proposals.[34] Like Dong, Gongsun constructed an etiological chain, rooted in the heart of the ruler, through which auspicious signs might once again be summoned. Gongsun argued that if good government were put into practice, the effect would be evident not only in the conditions of the emperor's subjects but in auspicious manifestations in the cosmos:

人主和德於上，百姓和合於下，故心和則氣和，氣和則形和，形和則聲和，聲和則天地之和應矣。

> If the ruler of men harmonizes his virtue to that above, and the Hundred Families harmonize in accord below, then their hearts will be harmonized, and so, *qi* will be harmonized. If *qi* is harmonized, then forms will be harmonized, and if forms are harmonized, then tones will be harmonized, and if tones are harmonized, then Heaven and Earth will resonate in harmony.[35]

During the initial stage, "the advent of harmony" (*he zhi zhi* 和之至), signs of cosmic regularity appear, including harmony between *yin* and *yang*, seasonable weather, and bountiful harvests. Gongsun also pointed to the absence of certain signs, such as mountains becoming denuded and marshes drying up. Gongsun posited a direct relationship between the virtue of the ruler, his ability to inspire a harmonious response from his subjects below, and the flourishing of the realm. Cosmic regularity implied governance in accord with Heaven.

Gongsun identified a more advanced stage, however, when "the ruler's virtue matches that of Heaven and Earth, bright as the sun and the moon" 德配天地, 明並日月. In this stage, "the ultimate harmony" (*he zhi ji* 和之極), there would be neither illness nor early death, and powerful manifestations of the ruler's cosmic virtue would appear—the unicorn and the phoenix, the River Chart and the Luo Writings. These auspicious emblems would suggest that the ruler's virtue had created a fully realized state of cosmic harmony:

> 則麟鳳至, 龜龍在郊, 河出圖, 洛出書, 遠方之君莫不說義, 奉幣而來朝, 此和之極也。

> Thus the unicorn and the phoenix arrive, the tortoise and the dragon dwell in the suburbs, and the River Chart and Luo Writings emerge,[36] so that none among the lords of distant lands fail to delight in duty or bear cloth tributes to court. This is the ultimate in harmony.[37]

For Gongsun Hong, a perfect symmetry existed between the appearance of auspicious and inauspicious signs and the presence of good or poor governance in the human realm. At the core of good governance were four basic virtues: humanity, duty, propriety, and knowledge. Together these constituted "the foundation of order and the application of the

Way" 治之本, 道之用也.³⁸ As he concluded his memorial, Gongsun explained away the baleful signs that appeared under virtuous rulers as residual effects of the actions of their predecessors. Heaven was both a model to be followed and an agent, possessed of power of its own, that demanded obedience. Its sons, rulers in the human realm, who failed to follow it, did so at their peril. Gongsun acknowledged, in blaming a drought in the time of Tang on the tyrant Jie of Xia, that cruel and benighted governance produced baleful signs.³⁹ Yet great bounties awaited the king who brought harmony to his own heart and to the hearts of his subjects. Should the human world be put into perfect order, Heaven and Earth would respond in kind, producing agricultural plenty and bountiful auspicious signs. The dearth of such signs was the wellspring of rhetorical power for both Gongsun Hong and Dong Zhongshu.

Gongsun Hong and Dong Zhongshu understood the production of auspicious signs in a manner that resonates with contemporary and near-contemporary writings in the *Huainanzi*, the *Shiji*, and liturgical traditions, such as those represented in the "Nine Songs" (Jiu ge 九歌) of the *Chuci*. Their memorials applied widely circulating legends concerning the production of auspicious signs in the past to the circumstances of the present. Dong in particular went beyond offering mere guidance, leveraging his knowledge of ancient exemplars to advocate concrete administrative and ritual reforms at Emperor Wu's court. In later decades, Emperor Wu's court did indeed produce liturgical repertoire, the nineteen-song set "Songs for the Suburban Sacrifices" (Jiaosi ge 郊祀歌), which served both to commemorate and to evoke the same kinds of signs that Gongsun and Dong claimed the emperor might bring into being.⁴⁰

Both Dong Zhongshu and Gongsun Hong pointed to harmonizing or accumulating virtue as the first step toward creating conditions that would ultimately produce auspicious celestial signs. The liturgy created for Emperor Wu's court would work toward this goal through ritual means, casting the emperor as an ideal supplicant seeking the aid of Heaven. Two stanzas from the hymn "O Supreme Progenitor" serve to demonstrate this ritual action:

雲風靁電,
降甘露雨,
百姓蕃滋,
咸循厥緒。

Let clouds and wind, thunder and lightning,
Send down sweet dew and rain,
So that the Hundred Families prosper and multiply,
And all perform their inherited tasks . . .[41] (lines 9–12)

鐘鼓竽笙,
雲舞翔翔,
招搖靈旗,
九夷賓將。

Bells, drums, and woodwinds play,
As the cloud dancers soar.
Let the starry banners of the Holy Ones fly,[42]
And the Nine Tribes come as guests.[43] (lines 21–24)

The song ritually imagines and instantiates a world in which auspicious signs both celestial and human appear, with the emperor gaining not only the support of the heavens but also the fealty of far-off allies who might, for a lesser ruler, have been competitors or enemies. The ritual evokes, and perhaps serves to bring into being, a world in which the empire both reflects and reinforces a cosmic balance, allowing for both peace and prosperity, the clearest proofs of legitimate rule.

Both Gongsun and Dong echoed the language and assumptions of Emperor Wu's edicts. Their arguments responded to the absence of emblems of cosmic regularity in the world. The symmetry between Gongsun's memorial and the edict to which it responded was almost perfect; both the auspicious emblems and precedents he cited were taken from the edict itself. Signs, too, appeared in perfect symmetry with the character of the ruler. For Gongsun, no baleful sign could appear under a good ruler; his view did not account for omens such as the pheasant that appeared during the High Ancestor's Day of Sacrifice or the violent winds that blew in the time of King Cheng. Gongsun adopted an encouraging tone and assured the emperor that the precise signs he desired would appear should he cultivate good governance. Dong Zhongshu went further afield, using classical texts and precedents not mentioned in the edict itself, most notably the narrative of the "Great Oath" in which King Wu received auspicious emblems that warranted his own legitimacy. Dong invoked narratives of decline that are absent in Gongsun's memorial. While Gongsun's language was designed to please the emperor, Dong's rhetoric provoked anxiety, reminding the emperor of the fallen, hallowed houses of the distant past.

Crisis at Court and the Comet of 12 BCE

Writing more than a century later, Liu Xiang and Gu Yong presented memorials concerned with the appearance of a baleful celestial sign in 12 BCE, Halley's Comet. It was a time when the future of the empire had once again become uncertain. The court had built the Changling 昌陵 mausoleum at great expense, disturbing existing grave sites and angering local residents, only to abandon the project.[44] Relatives of the Dowager Wang 王 (d. 13 CE) had held the post of Marshal of State, one after another, since 33.[45] After twenty years on the throne, Emperor Cheng had finally produced two children, but both had died in infancy.[46] A string of baleful signs had occurred, including an eclipse on the first day of the year and the appearance of the spectacular comet. Recently developed cosmological theories suggested that the time was ripe for dynastic change. Liu and Gu interpreted the comet against the backdrop of these events, reading both the sign and the world. Though their political alliances differed, Gu Yong being an ally of the Wang family, Liu Xiang a longtime critic, their rhetorical strategies were similar. Both marshaled expert knowledge of the Classics and of recent history to support claims that the comportment of the ruler had prompted the sign's appearance and that the proper response to it could avert real political disaster.[47]

Gu Yong

Gu Yong's memorial presented an etiology of both auspicious and inauspicious signs comparable to the etiologies of Gongsun Hong and Dong Zhongshu. It directly compared the appearance of signs under Emperor Cheng with signs that had occurred prior to his reign, arguing for a limited possibility of eliminating baleful signs. Recent work by Liu Tseng-kuei 劉增貴 has shown that Gu's technical work on calendrical calculations reveals his conception of history to be largely cyclical. Specifically, Gu's Three Troubles (San nan 三難) theory suggested that a major crisis would occur in 2 CE, 105 years after the adoption of the Grand Inception Calendar (Taichu li 太初曆) in 104 BCE. Because Gu's views on calendrical matters implied that historical cycles were not necessarily subject to human control, his rhetorical emphasis on the agency of the ruler is all the more remarkable.[48] Gu was quite clear that Heaven might select another house to rule in the Liu clan's stead, and he urged Emperor Cheng to end his liaisons with low-ranking (or low-born) people, to avoid raising taxes, and to reduce his excessive expenditures while there was yet still time. When troubling signs occur, it is crucial, Gu Yong argued, that they be addressed while they are yet in an incipient state, for there comes a point when nothing can be done.[49]

Gu outlined conditions that define good governance and lead to the appearance of auspicious signs. Gu advocated concrete actions to improve the state of the economy: lowering taxes, reducing court expenditures, providing relief to areas affected by flooding, suspending travel restrictions to allow for migration, and sending imperial envoys to inspect conditions in various regions of the empire. So long as these actions were taken, the world would prosper:

則卦氣理效，五徵時序，百姓壽考，庶中蕃滋，符瑞並降，以昭保右。

The *qi* of the hexagrams will appear with regularity, the Five Proofs will succeed each other in season, the Hundred Families will enjoy long life, the many grasses will prosper and multiply, and auspicious tallies and emblems will descend together, so as to illuminate the protection and aid [he has received].[50]

In this halycon vision, the Five Proofs, likely the Many Proofs of the "Great Plan"—rain, sunshine, warmth, cold, and wind—come in the proper seasons.[51] Grasses grow, *qi* follows regular patterns. The people live long because there is enough to eat. The austerity of the court leads to the prosperity of the empire.

At the same time, debauchery at court leads to social disorder and the appearance of baleful signs. Gu listed indulgent conduct and wasteful consumption of scarce resources as actions that "go against Heaven and do violence to creatures" (*ni tian bao wu* 逆天暴物).[52] The failure to employ good officials, orders to banish good men from court, and resentment among the Hundred Families preceded the appearance of baleful signs. Gu treated Heaven as a personified force, describing its responses in emotional terms; it had been stirred to rage. Signs of a poor ruler were both general and specific, celestial and terrestrial. Even as eclipses occurred, strange creatures appeared. The Five Planets strayed from their courses and mountains crumbled. Comets sparkled and peasants starved. The list of baleful signs concluded: "The Hundred Families will be short-lived, and the myriad creatures will die young or be crippled" 百姓短折，萬物夭傷.[53] These final signs were the most inauspicious of all, for they directly harmed human beings and other living things for whose welfare the ruler was responsible.

Gu Yong warned that failing to properly respond to a series of baleful portents would lead to the end of the dynasty:

終不改寤，惡洽變備，不復譴告，更命有德。詩云：「乃眷西顧，此惟予宅。」

If in the end you do not awaken and change your ways, the ill effects will accumulate and the [preconditions for dynastic] change will be complete.[54] You will not be warned again, but the Charge will be transferred to one who has virtue. The *Odes* says: "And he looked unto the west / This would be our home!"[55]

Whereas Dong Zhongshu had once responded to a lack of auspicious signs indicating Emperor Wu's possession of Heaven's Charge, Gu argued that recent signs suggested that Emperor Cheng was on the verge of losing dynastic legitimacy, citing an ode that ominously recalls the Zhou conquest of Shang. In the past year, there had been an unexpected eclipse, an unusually intense meteor shower, and the magnificent comet, all of which demanded immediate action. Nonetheless, the ongoing appearance of these baleful signs showed that the dynasty's fate remained subject to change. When they ceased to appear, it would mean that the ruling clan had truly been forsaken.[56]

Gu Yong described the problems that beset the emperor and his empire as the result of gradual developments allowed to go on for too long. "Calamity arises from the minuscule and incipient, and treachery is born from that which is lightly regarded" 禍起細微，姦生所易.[57] The key was to address the signs and change course before the sprouts had fully matured. It was best to eliminate calamity while it remained "minuscule and incipient," and thus susceptible to remedy. Yet recent disasters showed that the emperor's problems were already far beyond that point. The chance to eliminate the baleful signs, in Gu's rhetoric, was limited; the Liu house perched on the brink of destruction. Soon, Gu warned Emperor Cheng, it would be too late to save the dynasty:

> 彗星，極異也，土精所生，流隕之應出於飢變之後，兵亂作矣，厥期不久，隆德積善，懼不克濟。

> The broom-star is the most extreme of anomalies, born from the essence of the soil. The response of falling stars follows the aberration of famine, spawning civil wars in turn.[58] The time is not far off, when no matter how great your virtue or your accumulated goodness, I fear you will not be able to survive.[59]

Gu Yong recommended immediate action. First, Emperor Cheng needed to duly honor and promote those loyal, talented, and earnest men who had been passed over and distance himself from the debauched officials who had taken advantage of his largess, particularly those serving in the women's quarters. Gu warned the emperor to maintain strict ritual decorum, avoiding at all costs eating and drinking in the houses of his ministers and concubines,[60] potential rivals to the Wang clan.

The means to resolving baleful celestial signs lay in exposing and negating the conditions at court that had brought them into being. Gu's rhetoric was powerful, however, not only because it rooted its recommendations in the etiology of baleful celestial signs, but also because it claimed that those actions must be taken immediately, lest the dynasty's last chance for preservation slip away. Gu concluded his memorial by emphasizing that both baleful signs and the calamities they portend *could* be resolved, but only if the emperor "fixed his heart on doing good and rid himself of those with wayward ambitions" 定心為善，捐忘邪志。[61] In contrast with the polite niceties of the opening of his memorial, the language of Gu's conclusion bordered on goading:

> 竊恐陛下公志未專，私好頗存，尚愛羣小，不肯為耳！

> I only fear that Your Majesty has yet to focus your aim on the broader good, and that your personal proclivities remain, so you will continue to keep and cherish your throng of petty favorites, and be unwilling to take action![62]

Gu's memorial was a call to action, a pointed attack on Emperor Cheng's closest companions, male and female, the people with whom he chose to eat and drink and escape the dreariness of life as the political and religious head of the empire. Gu's attacks were never directed against the Wang clan. Indeed, if his language, stripped of its initial veneer of polite and apologetic formality, became caustic, or even rude, it was perhaps because he enjoyed the protection of powerful allies. Gu Yong could attack Emperor Cheng's favorites in the rear palace with biting rhetoric precisely because their power to retaliate against him was limited by his own relationship with the Wang clan.

Liu Xiang

Writing two years later, Liu Xiang cited many of the same baleful signs in his memorial, implicitly directed against the Wang clan,[63] that Gu Yong had in his attack against Emperor Cheng's favorites. Liu drew upon a slew of historical examples, from both the distant past and the imperial period, to make his case. Though the Han dynasty was in dire straits, the emperor could save it if he took the right actions.

Liu Xiang began by showing that appeals to historical precedent had a long history as a means of admonishing rulers, preparing Emperor Cheng for a serious discourse on the dangers that befall dynasties. Liu cited a litany of precedents, including the decline and destruction of the Shang dynasty, that might serve as a mirror for Emperor Cheng. Liu then recounted major signs that had occurred from the time of Qin Shihuang through the time of Emperor Cheng's grandfather, Emperor Xuan. Liu implied that examining the historical appearances of signs illuminated the processes through which they are brought into being; it was clear enough what sort of rulers received Heaven's favor and what sort of rulers lost it. All manner of baleful signs remained subject to elimination. Inauspicious omens had once been visited upon the courts of King Wuding of Shang and King Cheng of Zhou, both great rulers in antiquity. It was their exemplary responses to those signs that Emperor Cheng needed to emulate, for if he did, auspicious signs were sure to follow: "The response of the spirits is like a shadow or an echo, the whole world hears it in the same way" 神明之應, 應若景嚮, 世所同聞也.[64] Liu Xiang thus tempered his stern warning to Emperor Cheng with a reminder that he too might become a sage-king.

Liu Xiang pointed to the contingency of baleful signs, warning Emperor Cheng of the unprecedented number of such signs that had occurred during his own reign. Liu recalled the frequent eclipses in the time of Duke Xiang 襄 of Lu (r. 572–542), noting that one had occurred approximately every three years and five months during his reign. Early in the Western Han, under Emperor Jing, eclipses had occurred on average every three years and one month. But since the beginning of Emperor Cheng's reign, eclipses occurred with still greater frequency, Liu claimed, about once every two and a half years: "From ancient times to the present day, seldom has there been such a thing" 古今罕有.[65] Liu emphasized both the frequency and severity of recent signs, most notably the appearance of Halley's Comet two years earlier,[66] and the potential to eliminate them:

> 臣幸得託末屬，誠見陛下有寬明之德。冀銷大異，而興高宗、成王之聲，以崇劉氏。故狠狠數奸死亡之誅。今日食尤屢，星孛東井、攝提，炎及紫宮，有識長老莫不震動，此變之大者也。

> Your minister has been so fortunate as to belong to your clan, and I truly see that Your Majesty is possessed of broad and bright virtue. You hope to eliminate these great anomalies and gain a reputation like that of the High Ancestor and King Cheng, so as to honor the Liu clan. Thus, I have been willing to speak forthrightly, though I risk punishment by death. Now, eclipses of the sun are especially numerous, there has been a comet in Eastern Well and Assistant Conductors, with its fiery glow reaching Purple Palace,[67] and none among the aged men of knowledge does not quake with fear. And these are only the greatest of the aberrations . . .[68]

Much like Du Ye did in the case of the eclipse of 2 BCE,[69] Liu recalls the exemplars of the High Ancestor of Shang and King Cheng of Zhou, two figures who showed that baleful signs could be transformed into good fortune with the proper response. Liu effectively combines a technically detailed description of the comet's path, sources of classical authority, and an appeal to shared identity in the royal clan to convey both his own sincerity and the gravity of the circumstances the empire faced.

Without openly blaming any faction or group, Liu asked for a private audience to explain the implications of the signs using the technical charts in his charge. In support of this request, he cited the *Changes* and the *Documents*:

事難一二記, 故《易》曰「書不盡言, 言不盡意」, 是以設卦指爻, 而復說義。《書》曰「伻來以圖」, 天文難以相曉, 臣雖圖上, 猶須口說, 然後可知。

It is difficult to record these events one by one. The *Changes* says, "Writing does not fully express words; words do not fully express intent."[70] This is why, after one has set forth the hexagrams and indicated the relevant lines, one goes on to explain their meaning. Where the *Documents* says, "I sent them to bring the charts,"[71] the celestial patterns were indeed difficult to understand, for even once ministers have charted these for the ruler, it remains necessary to explain them orally, after which they can be understood.[72]

Liu got his private audience, but Emperor Cheng did not adopt his recommendations. The *Hanshu* tells us that each time Liu was summoned, he discussed the weakened position of the Liu clan and the rise in power of the Wang clan. Despite the memorial's elaborate account of signs past and present, it included no precise recommendations for changes to administrative policy. Liu reserved these, it seems, for his private meeting with the emperor. Gu Yong was able to employ vituperative rhetoric on account of his relationship with the Wangs, while Liu Xiang, for his part, dared not criticize them in open court.

Both Gu Yong and Liu Xiang wrote at a time when dynastic collapse appeared to be a real possibility—the empire was strapped for cash, imperial favorites dominated power, and the emperor had yet to produce or nominate an heir. The signs of danger had been numerous and persistent. It was not any single, isolated celestial sign that concerned them. Many baleful signs—celestial, terrestrial, and human—were woven into a broader image of the world. Both Gu and Liu examined the signs of the past two decades against what they knew of signs in the more distant past, what they knew of the historical events that followed, and what they knew of the court and the empire in their own historical moment. Both concluded that Emperor Cheng and the Han Dynasty had reached a treacherous precipice. Though their precise political alliances differed—Gu Yong an ally of the Wang family, Liu Xiang their longtime critic—the rhetorical strategies the two writers employed overlapped considerably. Gu Yong and Liu Xiang marshaled their expert knowledge of the Classics and of recent history to support claims that the appearance of baleful signs could be traced to the personal comportment of the ruler and that the proper response to such signs could avert real political disaster. The combined force of their rhetoric perhaps contributed to Emperor Cheng's decision the year before his death (8 BCE) to reform his Executive Council and to nominate his nephew,[73] the future Emperor Ai, as his heir. It was only after Emperor Ai's untimely death six years later that Wang Mang would take power as regent.

The Earthquake of 133 CE as a Celestial Sign

The earthquake of 133, like Halley's Comet in 12 BCE, came in the wake of troubling developments both at court and in the empire. In 132, military woes, ranging from Xianbei raids in Liaodong 遼東 to the rebellion of the "Monstrous Bandit" 妖賊 Zhang He 章河 (fl. 131–132) in Yangzhou 楊州, had beset the seventeen-year-old emperor.[74] Drought in the capital had led to famine; rituals to summon rain had been performed without avail.[75] Liang Na, whose brother Liang Ji 梁冀 (d. 159) would come to dominate the court from 142 until shortly before his death, had been made empress.[76] Counter to precedent, proposals had been made to enfeoff both Liang Ji and the emperor's former nursemaid, Song E 宋娥 (fl. 125–137).[77] Concerns about the poor quality of men nominated as "Filial and Incorrupt" had prompted a controversial order emphasizing literary capability and ability to cite the Classics.[78] Following the earthquake,[79] Emperor Shun issued an edict seeking explanations as to why it had occurred and what should be done about it. Both the edict and the three extant responses to it read the sign of the earthquake in view of other signs surrounding it:

詔曰：「朕以不德，統奉洪業，無以承順乾坤，協和陰陽，災眚屢見，咎徵仍彰。群公卿士將何以匡輔朕之不逮，奉答災異？災異不空設，必有所應，其各舉敦樸之士一人，直言厥咎，靡有所諱。」

The edict said: "I, despite my lack of virtue, serve a grand enterprise, but I have no means to receive and conform to Qian and Kun and bring *yin* and *yang* into harmonious order. Disasters and pestilences have repeatedly appeared, and proofs of blame continue to manifest. Nobles, ministers, and civil and military officials, how can you rectify and assist me in my carelessness, so that I can reverently respond to disasters and anomalies? Disasters and anomalies are not sent forth in vain, but there must be something to which they respond. Would that each of you nominate one honest and simple gentleman, to speak in a straightforward manner of the signs of blame, considering no need to censor anything."[80]

Common sense may suggest that an earthquake was a terrestrial rather than a celestial sign, but Emperor Shun's edict, like other writings on omenology, made no such division.

The earthquake was treated as a sign that issued from the ruler's lack of virtue, his failure to conform to and maintain harmonious cosmic patterns. The memorials of Zhang Heng, Ma Rong, and Li Gu conventionally addressed the earthquake as if it were a celestial sign, employing familiar arguments concerning etiology and elimination. Celestial, terrestrial, and human signs were part of the same world, bound together by resonance. Zhang Heng connected the earthquake to thunder and lightning, arguing that changes in the method of selecting Filial and Incorrupt men should be revoked. Ma Rong read the earthquake against agricultural and economic conditions, arguing for increased official attention to the seasonal needs of farmers. Li Gu, whose striking memorial carried the day, treated the earthquake as a warning from Heaven requiring the disenfeoffment of the emperor's former nursemaid and greater circumspection in filling official positions. The three memorialists—much like Gu Yong and Liu Xiang—responded to the manifold political, economic, and cosmological circumstances that confronted them, not to an isolated sign with a fixed meaning.

ZHANG HENG

Zhang Heng's technical acumen and official position made him extraordinarily qualified as a reader of celestial signs. Just as Sima Qian had two centuries earlier, Zhang Heng held the post of Senior Archivist and, in addition to his works on history, he had produced star charts, composed a treatise on celestial patterns, and made improvements to the armillary sphere, adding rings for the prime meridian and the horizon.[81] He was an expert in various sign-reading techniques, including the reading of the wind, and an observant historian who both criticized the apocrypha for historical inaccuracies[82] and identified multiple points of disagreement between the *Shiji* and *Hanshu*.[83] Though his system of astronomical computation was never adopted, Zhang's readings of signs must have carried great weight at Emperor Shun's court. In 132, Zhang successfully developed a seismoscope sensitive enough to identify distant earthquakes that could not be felt in the capital.[84] The name given to the seismograph, the "Instrument for Observing Wind and Movements of the Earth" (*Hou feng di dong yi* 候風地動儀),[85] underscores the continuity between the earth and the atmosphere, the terrestrial and the celestial. The innovation was significant enough to be recorded not only in Zhang's biography but in the "Basic Annals" for the reign of Emperor Shun as well.[86]

In the opening of his memorial, Zhang Heng treated the 133 earthquake as if it were in the same category as violent weather manifestations. Zhang began with a standard etiological claim: "I have heard that when there is good governance auspicious signs descend, and when there is repugnant governance, proofs of blame appear" 臣聞政善則休祥降, 政惡則咎徵見.[87] Zhang Heng referred not only to the earthquake in his memorial but to recent weather manifestations that had appeared in conjunction with it:

間者, 京師地震, 雷電赫怒。夫動靜無常, 變改正道, 則有奔雷土裂之異。

Recently, in the capital there has been an earthquake, and thunder and lightning have blazed in anger. When movement and stillness are without constancy, and aberrations alter the correct way, then there are anomalies consisting in rolling thunder and cracked soil.[88]

Signs above and below, Zhang Heng claimed, constituted warnings from Heaven: "Monstrous stars appear above; quaking and splitting manifest below. Heaven's warning is clear. It is enough to chill the heart" 妖星見於上, 震裂著於下, 天誡詳矣, 可為寒心.[89] Zhang cast both the earthquake and the stormy weather as warnings issuing from a common celestial source.

Zhang asserted that those signs came about due to recent changes in policies for nominating individuals as Filial and Incorrupt. The new nomination procedures, adopted in the eleventh month of the previous year, required all but the most talented candidates be at least forty years of age, and moreover, the new rules required mastery of the language of the Classics and skill in composition.⁹⁰ Though himself a brilliant rhetorician and poet, Zhang held that this emphasis on mastery of language overshadowed the deeper ethical values of classical learning. Appealing to a well-known exemplar of filial piety, Kongzi's unrefined and dull (*lu dun* 魯鈍) disciple, Zengzi 曾子, Zhang held that the most filial candidates might be ineligible for selection under the new rules.⁹¹ The new system, he warned, made it difficult to distinguish the truly virtuous from the merely eloquent. Under such conditions, it was no wonder earthquakes and inclement weather occurred:

今真偽渾淆，昏亂清朝，此為上陵下替，分威共德，災異之興，不亦宜乎？

Now, the true and the false are confounded, and darkness and disorder are visited upon the Clear Court. Because of this, those above fall into decline and those below are abandoned, dividing up authority and sharing power. As for the frequent appearance of disasters and anomalies, is it not fitting?"⁹²

The distinction between good and poor officials had grown vague, and the distinction between ministers and the ruler had been eroded. Ministers arrogated the authority and capacity to distribute favors that properly belonged to the ruler alone. The earthquake and concurrent violent weather manifestations could be traced to political conditions at Emperor Shun's court and remained subject to change. Zhang opened and closed his memorial with reminders that baleful signs did not always or inevitably lead to calamity. In his preamble, Zhang appealed to the classical example of King Cheng's response to the violent winds.⁹³ In his final line, Zhang proclaimed that while the time for action was long overdue, reform might still avert disaster: "Those who are clear-sighted resolve calamity before it sprouts. Now, the sprouts have already appeared, but if you refine your governance, trembling with fear, you will transform calamity into blessing" 明者消禍於未萌，今既見矣，修政恐懼，則轉禍為福矣.⁹⁴ Crisis could yet be averted but had to be averted soon.

Ma Rong

Ma Rong's response to the earthquake was less a reading of the sign itself than a reading of the state of the economy. Ma never once referred directly to the quake. The earthquake was a sign, but a comet, an eclipse, or any other event that could be treated as a warning from on high might have served just as well. Ma claimed in his opening that so long as the empire was properly ordered, auspicious signs of celestial constancy would manifest themselves in the heavens. Following the order inherent in the heavens served to maintain that order, bringing wind and rain at the right time and leading to bountiful harvests. But when the celestial order was lost, famine appeared as an inauspicious sign.[95] The advent of famine was the sign that most concerned Ma Rong, and it became the central object of his economically oriented admonitions.

Ma invoked a set of regulations for following Heaven, including ordinances directing which activities should be carried out at which time of year. Though those regulations might be complete and perfectly articulated, imperfection in their enforcement was the source of baleful signs in Heaven and in society:

天猶有不平之效，民猶有咨嗟之怨者，百姓屢聞恩澤之聲，而未見惠和之實也。

In the heavens, there continue to be events that verify that all is not tranquil, and among the people, there continue to be sighs of resentment. The Hundred Families have often heard talk of grace and favor, but have yet to see the substance of benefits and harmony.[96]

The crux of the problem, for Ma, lay in the disjunction between substance and rhetoric. Where Zhang Heng was satisfied to speak of baleful signs as shadows and echoes, Ma insisted that imperial grace must be more than talk. The crucial signs implied a bountiful harvest. Famine and starvation, the most inauspicious of signs, signaled visible and present suffering among the people.

It was not enough, in Ma's view, to announce the virtue of the emperor through the production of regulations. It was the duty of the imperial administration to enforce regulations in accord with the balanced regularity of the heavens. Under the present conditions, however, "excessive severity" had come to be the norm:

今從政者變忽法度，以殺戮威刑為能賢。問其國守相及令長何如？其稱之也曰「太急」，其毀之也曰「太緩」。

Now, those who carry out the administration alter and disorder the laws and measures, so that those who engage in slaughter, intimidation, and punishment are regarded as capable and worthy. If we inquire about governors, chancellors, or local magistrates, those who praise them say they are excessively severe, but those who condemn them say they are excessively lenient.[97]

Ma argued for moderation in the enforcement of regulations. Though he advocated more lenient policies, he emphasized that neither excessive severity nor excessive leniency should be tolerated. Both threatened the cosmic balance. The tendency to favor severity over leniency lay at the root of recent disasters: "Severity brings on cold, and leniency brings on heat. Both are equally wrong,[98] yet those who discuss them forgive severity. This is why *yin* and *yang* are out of balance" 夫急致寒, 緩致燠, 二者罪同, 而論者許急, 此陰陽所以不和也.[99] Understanding the etiology of the disaster implied the method for its elimination.

Ma claimed that the emperor could restore cosmic regularity by balancing leniency and severity.[100] Instituting kingly governance (*wang zheng* 王政) required punishing officials who practiced excessive severity and those who practiced excessive leniency in the same way. Cosmological balance required balance in the implementation of policy:

好惡既明, 則宰官之吏, 知所避就。又正身以先之, 嚴以涖之, 不變則刑罰之。夫知為善之必利, 為惡之必害, 孰能不化?

Once good and bad have been clarified, the various officials will know what to avoid and what to follow. Moreover, rectify yourself and lead them, sternly watch over them, and if they do not change, then punish them with the laws. For if they know that doing good will certainly profit them, and that doing ill will certainly harm them, who will fail to be transformed?[101]

Both the origins and resolution of the cosmological problems facing the empire lay in the political realm. Officials meting out punishments with excessive severity led to the imbalance in *yin* and *yang* and might have produced a host of signs, including earthquakes, floods, and droughts. If, faced with punishment, officials came to realize that excessive severity was as damaging as excessive leniency, both the cosmological problem and the political problem would be resolved.

Ma Rong invoked cosmology but continually tied cosmic imbalances to agricultural production. Appealing to the Classics, Ma suggested that agriculture was the first responsibility in governance and implied that military unrest in various parts of the empire resulted from a failure to encourage agricultural production:

臣聞洪範八政，以食為首；周禮九職，以農為本。民失耕桑，饑寒并至，盜賊之原所由起也。

I have heard that in the Eight Concerns of Governance in the "Great Plan" food comes first, and in the Nine Official Responsibilities in the *Rites of Zhou*[102] farming is considered the most fundamental. When the people fail to plow or attend to silk production, then starvation and cold both come upon them. This is the path by which thieves and bandits arise![103]

Ma recalled an ancient period when the economic needs of the people had been better fulfilled. The people themselves had been frugal and assiduous, held austere weddings and funerals, and worked hard. Most importantly, however, officials did not interrupt agricultural production. The prevalence of banditry in his own time was the result of spotty law enforcement, artisans producing useless implements, merchants trading in rare goods, and farmers failing to perform agricultural tasks. Famine was a consequence of an economic system in which there were easier ways to make money than producing plows, selling grain, and working the land.[104]

Ma's approach to the earthquake of 133 centered on economic troubles rather than the earthquake itself. The earthquake was a cosmic symptom of problems in the world of human affairs. The solution lay in correcting the unbalanced implementation of official policy. Excessive severity had been valued despite its damaging effects, taking farmers away from their work, and thereby producing famines and other economic disturbances. Ma Rong did not blame any person or faction at court, and claimed that workable regulations and seasonable ordinances were already on the books. Resolution of both economic and cosmological problems lay in ensuring that those regulations and ordinances were properly enforced.

Li Gu

Emperor Shun selected Li Gu's memorial as the finest he received, perhaps due to its simple and concrete advice. Li argued that the emperor's former nursemaid, Song E, should not be ennobled, for there was no precedent for such an action. Li treated the body of the emperor as a microcosm of the empire and analogized good governance to well-maintained dikes able to stem any flood.

Li began with a typical description of the etiology of baleful signs, but this commonplace was by no means perfunctory. By claiming that the earthquake was a warning from Heaven, he implied that the ennoblement of Song E violated Heaven's will. Li knew to tread lightly, and he carefully avoided insulting Emperor Shun's childhood favorite. Instead, he cited the absence of support in the Classics for raising the status of a nursemaid to that of an aristocrat:

> 阿保雖有大功, 勤勞之恩, 可賜以貨賄, 傳之子孫, 列土分爵, 實非天意。漢興以來, 賢君相繼, 豈無保乳之養? 非不寵貴之。然上畏天威, 俯察經典, 不可, 故不封也。

> Your nursemaid, Song E, certainly has a great deal of merit, and the grace of her diligent labor might be rewarded with goods, but to vest her with a heritable property, land, and a title, truly is not Heaven's will. Since the Han arose, worthy rulers have succeeded one after another. Who among them has not been nurtured by a nursemaid? It is not that they did not cherish and honor [their nursemaids]. Yet, when they looked up, they feared Heaven's majesty, and when they looked down and examined classical texts, [they found that] it was not permissible, and so they did not enfeoff [their nursemaids].[105]

Li Gu linked excessive grants of rewards and authority to the emperor's favorites with the appearance of baleful signs. Heaven issued the earthquake and other recent signs to warn Emperor Shun against doling out inappropriate favors.

The body played a crucial role in Li Gu's rhetoric. Li used the body in three ways. First, while he focused on a narrow, precise source of baleful signs, Li identified the body of the emperor with the whole of the realm, so that earthquakes and eclipses appeared in the body of the emperor himself:

王者父天母地，體具其山川。今日蝕地動，山崩晝晦，主將安立？物將安寄？

The king takes Heaven as his father and Earth as his mother. His body contains their mountains and rivers. Now there are solar eclipses and earthquakes, mountains crumble, and it grows dark in the light of day. Where can the ruler stand in repose? When can his creatures lodge themselves in repose?[106]

Emperor and empire were of a single body. Second, Li used the rhetoric of the body to state the critical importance of employing forthright (*fang zhi* 方直) ministers and rejecting those who were wayward and glib (*xie ning* 邪佞). The word *ti* 體 in its basic sense referred to the body or the limbs, and as a verb it meant to be on intimate terms with another person. Li used the phrase *xiang ti* 相體 (to be on intimate terms with one another) to describe the ideal relationship between ministers and rulers. He asserted that if His Majesty raised up loyal and good (*zhong shan* 忠善) men, then "lord and minister would be on intimate terms with one another, and the relations between the ruler and his subordinates would flourish" 君臣相體，上下交泰.[107] Close relationships implied more than impersonal court etiquette.

Briefly leaving aside the image of the body, Li next analogized the administration as a series of dikes along a river. So long as they maintained their integrity, they would not break even if heavy rains fell. So long as governance was based on moral suasion (*zheng jiao* 政教), a year of famine would not threaten the empire. However, Li warned, if the dikes sprung leaks it might not be possible to repair them, and if the wise and the worthy (*xian zhi* 賢智) fled the court, it might not be possible to bring them back. Tempering his language, Li warned the ruler of the current state of things: "The dikes remain solid, but holes are gradually appearing" 隄防雖堅，漸有孔穴.[108]

As did Gu Yong, Li suggested that while baleful signs could still be addressed, a point might come in the future when nothing could be done. The ruler could resolve baleful signs but only up to that point. He needed to act before the dikes began to leak, for after leaks had sprung it would be too late to stem the flood. Once moral suasion was no longer effective and the last capable and loyal men had withdrawn from the court, good governance would no longer be conceivable. Li then shifted from the analogy of the dike back to that of the body:

> 譬之一人之身，本朝者，心腹也；州郡者，四支也。心腹痛則四支不舉，故臣所憂，在腹心之疾，非四支之患。

> Compare this to the body of a man. The court is the heart and belly, while the provinces and commanderies are the four limbs. When the heart and belly ache in pain, one cannot lift the four limbs. What I am worried about is the illness of the heart and belly, not the troubles of the four limbs.[109]

Li recommended a single minor action, canceling the enfeoffment of Song E. By evoking an image in which the court was the heart and torso of the empire, while the provinces and commanderies were the four limbs, Li reinforced the sense that problems in the central court had powerful ripple effects. Rectifying the court would bring prosperity and security to the entire empire. Li continued:

> 臣以為堅隄防，務政教，先安心腹，釐理本朝，雖有寇賊、水旱之變，不足介意也。令隄防壞漏陋，心腹有疾，雖無水旱之災，天下固不可不憂矣。

> I believe that to make solid the dikes and give priority to devising civilizing policies, one must first bring repose to the heart and belly, bringing order to the court. Even if there are raiders and bandits, and the aberrations of floods and droughts, these will be no cause for concern. But if the dikes break down and water leaks through, if the heart and torso grow ill, then even if there are no disastrous floods and droughts, it is certain that [no one in] the realm will be without cause for worry.[110]

Li called into question the significance of the signs, be they drought and floods, raiders and bandits, or earthquakes and eclipses. The only security lay in setting the court to order, making the dikes steadfast, ensuring the health of the heart and torso. Even if baleful signs did not appear, a court in disorder would lead to calamity. Conversely, a well-ordered court that ruled by moral suasion would maintain its integrity whatever the challenges it faced.

The responses of Zhang Heng, Ma Rong, and Li Gu to the earthquake of 133 show that no simple, straightforward system of correspondences between specific signs and specific events commanded authority. Such a rigid system would have stripped memorialists of a great source of rhetorical power. Despite the sense of system and order treatises on celestial signs had created, memorialists could interpret each sign as the product of its own unique circumstances. Memorialists drew freely on metaphors, precedents and principles, authoritative texts, and cosmological correspondences, with no hard and fast rules for interpreting signs.

Conclusion

Interpreters of celestial signs read the contours of the world in which they lived, including its natural, ritual, and political aspects. They examined the shape of the economy, power structures at court, diplomatic and military engagements, and the textual traditions at their disposal, including technical treatises, historical records, and the Classics. Ominous as the anomalies might have been, the most skilled interpreters cast the signs themselves as less important than the conditions in the human world in which they appeared. As Yang Xiong wrote, "the sages take virtuous action to be their chief task, treating anomalies as no more than secondary concerns" 聖人德之為事，異亞之.[111] Good governance was paramount.

While rhetorical practice surrounding baleful signs changed over time, concern with the relationship between signs and the human world was constant. Dong Zhongshu and Gongsun Hong, early in the reign of Emperor Wu, expressed greater concern about the absence of auspicious signs than about the presence of baleful ones. Later in the Western Han, memorials invoking disasters and anomalies became commonplace. Prompted as much by perceived dynastic instability as by the comet of 12 BCE, Liu Xiang and Gu Yong cited sign after inauspicious sign as they warned Emperor Cheng to get his house in order or risk dynastic collapse. Citing baleful signs and asserting how they had come into being and what might be done about them had become an established practice. In the mid-to-late Eastern Han, even as technical experts such as Zhang Heng developed new technologies for identifying celestial signs, including improvements to the armillary sphere and building a seismoscope, claims regarding the meaning of any sign could only be made with reference to the human world. For Zhang Heng, the earthquake of 133 was a sign that the court was recruiting officials in the wrong way. For Ma Rong, it was a sign that officials were excessively severe in their administration. For Li Gu, it was a sign that Emperor Shun had shown inappropriate largess to his favorites. Memorialists detailed the etiology of baleful signs and claimed that if the conditions that produced the signs were removed, so too would be the events they presaged. What concerned memorialists most were the inauspicious actions and events they saw in the human world.

Conclusion

Interwoven Discourses and Contextualized Interpretation

Urging scholars to take up the dual threads of nature and culture and "retie the Gordian Knot," Bruno Latour acknowledged the potential for resistance: "That a delicate shuttle should have woven together the heavens, industry, text, souls and moral law—this remains uncanny, unthinkable, unseemly."[1] Celestial signs were at once natural phenomena and cultural ones. This book has positioned the work of those who observed and interpreted them within Lloyd and Sivin's cultural manifold and has presented celestial signs as part of a world in which nature and culture, science and religion,[2] the heavens and the human realm, were inextricably bound. It has followed several threads, traversing the open range of academic territory, trespassing boundaries where it has been necessary or beneficial to do so, examining technical manuals and historical narratives, court ritual and the administration of the empire. Multiple types of texts existed in the same textual communities, and in many cases, the same individuals engaged celestial signs in a variety of discourses.[3] Interpreting those signs involved reading circumstances at court and in the empire, identifying past instances of similar signs, and developing cosmological theories. It was not a matter of simply making one-to-one correspondences between signs and events but of positioning signs in the meaning-laden world that surrounded them.

Classical Authority in Technical Writing and Memorials

Following the Qin unification of a handful of formerly distinct states, the two Han dynasties were an epoch in which disparate types of texts were collected and brought to bear upon one another. The imperial court, regional kings, and high-ranking officials such as Li Cang collected caches of texts that allowed a privileged and talented few to gain mastery of a vast array of subjects. The cache of texts discovered at Mawangdui offers a glimpse into one textual community at a regional court in the early Western Han. Among the texts available in that community were technical manuals that identified specific correspondences between signs and events, as well as works resembling both Classics and masterworks (with no apparent generic distinction between them), including the *Constant Models* and materials attached to the *Changes*, that dealt with the general role sign-reading played in statecraft. These described the duty of the ruler to emulate the workings of Heaven and provided a cosmological basis for personal and political administration. The presence of both types of texts in the same communities made it possible for readers and thinkers to consider one in the light of the other, even if individual tomb texts themselves did not cite or refer to one another. In the decades that followed, the compilers of the *Huainanzi* and the *Shiji* edited, organized, and arranged formerly discrete types of texts into large-scale textual projects. The *Huainanzi* and the *Shiji* aimed for both universality and coherence, even if they did not always achieve it. In them, narratives surrounding exemplary figures, cosmological theories, and technical information on sign-reading were all brought to bear on one another. In the late Western Han, the court sponsored a project to collect, edit, and collate texts from throughout the empire. Liu Xiang and his son Liu Xin cataloged the texts gathered in the imperial library and composed works that—in all probability—formed the basis for the *Hanshu* "Wuxing Treatise." Yang Xiong, who had worked alongside the Lius in their bibliographical endeavors, promoted the notion of the Five Classics as a unitary corpus of the highest value.

By the Eastern Han, when the received versions of the *Hanshu* technical treatises were compiled, sign-reading had become inseparable from the authority of the Classics. The "Wuxing Treatise" drew its basic structure from the "Great Plan" chapter of the *Documents* and served as a kind of *Annals* for Han times. The "Celestial Patterns," while largely drawn verbatim from the corresponding treatise in the *Shiji*, distinguished itself from its predecessor by integrating copious citations from the Classics into new technical passages and by including a chronological account of celestial signs in the Western Han modeled on the *Annals*. Both developments show the ways in which conventional textual practices such as the keeping of chronological court records and the selection of relevant classical citations were hybridized with technical manuals for sign-reading.

Memorialists in the late Western Han and Eastern Han likewise employed both technical knowledge and direct citation of the Classics to lend rhetorical power to their interpretations of celestial signs. At the outset of Emperor Wu's reign, memorials in the biography of Dong Zhongshu that argued for decreased use of harshness, increased use of moral suasion, and the establishment of new educational institutions cited the *Documents*, the *Odes*, and the *Annals*. But Dong's dynamic use of the Classics remained somewhat exceptional in his time, if the language attributed to him was in fact his own.[4] His contemporary Gongsun Hong, who like Dong was an expert in the *Gongyang* tradition of the *Annals*, eschewed direct citation of the Classics in his own memorials on the absence of auspicious celestial signs during Emperor Wu's inaugural years. Dong's use of the Classics anticipated a rhetorical mode that would become commonplace in the late Western Han and the Eastern Han. Indeed, he was the only official active prior to the reign of Emperor Yuan to be cited with any frequency in the *Hanshu* "Wuxing Treatise," and even there his views are often conflated with Liu Xiang's. From the late Western Han on, however, memorialists routinely wove citations from the "Great Plan" and the "Appended Statements" into arguments concerning why baleful signs appeared and how to eliminate them. The old view, held by Duke Jing of Qi, that baleful signs appeared due to random misfortune, had largely disappeared; like Master Yan, memorialists nearly always claimed that signs appeared for identifiable reasons. They appealed to narratives concerning exemplary figures properly responding to such signs, found both in the Classics themselves and in other historical texts. Erudite officials peppered their memorials on celestial signs with language drawn from the Classics, advancing arguments that reflected their own concerns and their understandings of the needs of the emperor and the empire. Knowledge of historical and legendary instances of signs allowed memorialists to compare past events with present circumstances and advance claims concerning how ideal rulers respond to bad omens. Though many oft-cited signs dating to the Chunqiu period and earlier were recorded in the Classics, memorialists also appealed to Warring States and Qin-Han era events transmitted in masterworks and historical texts. Rhetorically adept memorialists marshaled signs past and present, paired with broad textual knowledge, to advance their views concerning how best to address the problems the empire faced.

The most learned and versatile writers operated in multiple genres. Liu Xiang edited classical texts, compiled an omenological treatise, and produced memorials that were dazzling displays of both technical knowledge and classical learning. Ma Rong helped Ban Zhao to compile the "Celestial Patterns" treatise and displayed hard-headed practicality buttressed by classical learning in his economic arguments at court. Ma's textual expertise regarding Classics, masterworks, and poetry perhaps even exceeded his technical knowledge. Ma wrote notes on the "Li sao," the *Laozi*, the *Huainanzi*, and the *Lienü zhuan* 列女傳 (Biographies of Exemplary Women) and commentaries on the *Analects*, the *Odes*, the *Documents*, the *Changes*, the *Record of Rites*, and the *Rites of Zhou*.[5] Zhang Heng, the consummate scientific thinker, engineer, and poet, wrote astronomical treatises, invented the seismoscope and improved the armillary sphere, and engaged in governance and politics. Each of these writers, it must also be noted, applied his expertise regarding both the Classics and celestial signs to composing celestial journey verse. Liu wrote of receiving a hearing among the stars where he protested having been misunderstood and mistreated at the human court. Ma integrated classical citations with celestial journey verse to argue for the necessity of maintaining martial valor and military preparedness. Zhang composed a celestial journey poem that inverted fundamental conventions of the genre, elevating the journey through classical texts over the journey through the stars.[6] The methods and materials required for engaging with the development of celestial journey verse differ substantially from those that fit within a study of omenology. Notwithstanding, celestial journey verse remains a promising avenue for research into the broader valences of celestial signs (in the more general semiotic sense, rather than the specific sense of omens) and their relationship to classical authority.

While modern disciplinary divides may leave contemporary scholars inclined to variably view such figures as poets, or statesmen, or technical experts, they were in fact all of these at once. Long periods of (relative) stability in Han times created the possibility for both the imperial court and regional centers to compile vast collections of texts in a multitude of genres. Those with the erudition and opportunity to work directly with these texts drew connections between them. Their knowledge of the Classics and masterworks, of history, and of technology, married with their own experiences at court and in service to the empire, formed a totality, partially shared and partially unique to each individual, that made up their view of the world.

Sign-Reading as Reading the World

The interpretation of signs required understanding the relationship of a given sign to the social and cosmic dynamics of the world around it. Reading signs involved reading a host of factors; the relationship between signs and events emerged against the backdrop of politics and history, the state of the economy and the environment. Sign-reading identified the current trajectory of events rather than predicting the future. There was no notion of fate in the Western sense, no inevitable tragedies bound to happen no matter how the actors involved behaved.[7] The "Essentials" manuscript from Mawangdui showed that even the best readers of signs could produce accurate prognostications no more than 70 percent of the time. To the extent that signs predicted misfortune, the failure of such events coming to pass confirms rather than refutes arguments concerning etiology and elimination. Time and again, memorialists claimed that removing the conditions that had produced a given sign would eliminate the sign itself and guard against future misfortune.

Technical manuals for sign-reading were, moreover, one authority among many. The Mawangdui tombs contained many different types of texts that may have commanded authority when a baleful celestial sign appeared. These included not only technical texts but also texts dealing with good governance and the relationship between the heavens and the human realm. Some would be lost, while others would later be classified as masterworks and Classics or incorporated piecemeal into large-scale Western Han textual projects. The *Shiji* and *Huainanzi* contained cosmological theories and historical narratives that might have been brought to bear on sign-reading. And in the late Western and Eastern Han, technical sign-reading texts were dependent on and intertwined with the authority of the Classics, especially the *Annals*, the "Great Plan," and the "Appended Statements." Memorials on baleful signs interpreted events and processes in the human world, primarily those concerning the life of the court and the administration of the empire. Appeals to cosmic processes mapped onto flesh-and-blood human affairs. Eclipses and earthquakes called for reading the dynamics of power within the palace as much (or more) than they called for reading abstracted dynamics of *yin* and *yang*. Indeed, it was in the human world that the interplay of cosmic forces was most fully and forcefully instantiated. Technical manuals presented guidelines for reading various types of signs, not absolute, exclusive rules.

Authoritative texts, cosmological systems, and historical precedents nonetheless constrained interpretation. Technical manuals offered possibilities for reading certain signs, facilitating linkages between them and specific types of events. While celestial signs were sometimes associated with deities, by the late Western Han, it had become common to explain their appearance through complex cosmological systems constructed around *yinyang* and *wuxing* dynamics. These did not entirely displace attributions of celestial signs to the willful actions of spirits, gods, or an anthropomorphic Heaven, but they did become the dominant mode of discussing the etiology of signs and how they might be eliminated.[8] Cosmological correspondences, historical narratives, and authoritative texts made up systems of meaning that were shared among members of communities. Adept sign-readers rooted their claims in these systems, drawing on an immense wellspring of rhetorical power.

Sign-reading always occurred within some human context. Viewing technical manuals in isolation might make it seem as if a given sign always pointed toward a specific type of event, and that the role of a manual's user was simply to identify the correct sign and match it with its correspondent. However, the historical record shows sign-readers acting in a far more sophisticated manner. Readers of celestial signs brought into their arguments citations from the Classics, historical records, evaluations of military, economic, and ritual policies, and analyses of palace life, power struggles, and the affairs of distaff families and the royal line. Memorialists directed their responses to baleful signs toward the myriad circumstances that surrounded them. The situation of the court and the empire at the time of the sign's appearance, not the precise technical details of the sign itself, became the focal point of their arguments. Reading celestial signs meant reading the state of the world at a particular moment in time.

Appendix A
Terminology and Translation

Translation is imperfect. In moving from one language to another, something is typically gained or lost, for words in different languages often do not neatly correspond to each other. Though we may search for hypothetical equivalents, once a given word comes into use in the language into which it is translated, it takes on a life of its own.[1] The technical vocabulary surrounding celestial signs is difficult for a number of reasons. First, many words might simply be rendered as "sign" or "omen" if they occurred in isolation. Second, words often have more than one sense, varying with context. Third, a standard of consistency must be maintained for those readers who cannot access the primary texts in the source language. Let us examine the renderings employed here in turn.

Zhan 占: Sign-reading (sometimes abbreviated as "reading")
    Zhan refers both to the process of conducting a ritual for reading signs and the information that results from it. Rendering *zhan* as "divination" or "divine" implies a consultation with a deity. Deities are sometimes involved in sign-reading, but often the process works through mechanisms that are either not clearly identified or are naturalistic, as in the case of the identification of imbalances in *yin* and *yang* or the *wuxing*. The rendering of *zhan* as "prognostication" is also problematic because it implies clear foreknowledge of fixed future events. Sign-reading in early China sometimes gestured toward knowledge of the future, but rarely implied that the future was fixed or determined. Instead, *zhan* seemed to identify the trajectory of events over time. This allowed readers of signs to forecast disastrous events that would occur *so long as the current trajectory remained unchanged*, as well as to take actions to ameliorate or avoid disastrous outcomes.

Xiang 象: Counterpart, image; sign

Xiang tend to be stable, visually recognizable signs. Because they often correspond to objects in the terrestrial world, Schafer refers to *xiang* as counterparts, a convention this volume generally follows.[2] *Xiang* are not normally disasters, omens, or deviations from the constant order of things. The Sixty-four Hexagrams of the *Changes* are *xiang*. The sun, moon, and stars are *tianxiang* 天象, celestial images, or to the extent that their correspondences with people and places in the human world are at play, celestial counterparts. NB: The sense of "sign" in celestial signs, as the phrase is used in the present work, is meant to encompass a much broader range of phenomena than *xiang*.

Rui 瑞: Auspicious emblems[3]
Fu 符: Tallies

Rui are the most auspicious of signs: the sweet dew that falls from the heavens, the unicorns that appear in the realm, the ancient tripods that emerge from a river.[4] Such events confirm the legitimacy of the sitting ruler. In its most literal sense, a *rui* is a jade tally that authorizes its possessor to perform certain actions, such as raising troops. The implication of the appearance of *rui* is that the ruler has been granted authority by Heaven, his ancestors, or the gods. *Fu* are documents that civil or military officials carried as signs of authority, typically made of wood or bamboo, sometimes split into two halves that could be joined to verify their authenticity. In religious contexts, *fu* could serve as talismans for protection against malevolent ghosts or demons, and often used language that mimicked a bureaucratic register.[5] In the present volume, *fu* occur as signs of divine authorization for a current ruler, as in the phrase "tally of having received the charge" (*shou ming zhi fu* 受命之符). Owing to their shared sense as signs of political legitimacy, *fu* and *rui* frequently appear as a pair, "auspicious tallies and emblems" (*fu rui* 符瑞).

Zai 災: Disaster
Yi 異: Anomaly

Zai and *yi* often appear together in the phrase *zaiyi* 災異 from the late Western Han onward.[6] They include many types of baleful omens: fires, floods, droughts, plagues of locusts, comets, eclipses, and more. *Zai* and *yi* are particularly prominent as ways of referring to baleful omens in the traditions surrounding the *Annals*. The *Gongyang Tradition* refers to *yi* some thirty-three times and *zai* sixteen times. The two terms refer to overlapping types of phenomena.[7]

Hua 化: Transform/transformation
Bian 變: Aberrate/aberration; alternate/alternation

Hua and *bian* both signify changes in state and may at first glance appear to be synonymous. However, their respective connotations are far different. *Hua* is a process through which the sage ruler and his court transform the empire and its subjects, creating a foundation for harmony and prosperity throughout the realm. *Bian*, in omenological discourse, signifies the appearance of a baleful sign, a deviation from the regular, constant state of things, and also sometimes specifically refers to an actual or potential change in the ruling house. In discourse surrounding the *Changes*, *bian* signifies the alternation of *yin* and *yang* lines[8] and has a neutral connotation.

Zheng 徵: Proof; verify
Xiao 效: Verifying event, verify
Yan 驗: Verify, verification

The terms *zheng* and *xiao* are largely synonymous in the texts under review here. Both terms refer to events that verify or corroborate the significance of earlier signs. Hence, the civil war that broke out in the capital in 91 BCE is labeled a *xiao*, a verifying event that confirms the interpretation of a sign that occurred two decades earlier: a comet appearing in the constellation Eastern Well in 110 BCE.[9] *Zheng* likewise signals that an earlier sign, or sometimes an earlier statement, is meaningful. When Zhang Heng's seismograph was triggered for the first time, officers in the capital complained there was no *zheng* (proof) until a messenger arrived, several days later, reporting an earthquake in the direction the seismograph indicated.[10] Baleful signs may also be *zheng*. Gu Yong pointed to proofs of blame (*jiu zheng* 咎徵) that were likely to occur should Emperor Cheng fail to follow his policy recommendations. These included eclipses, planets straying from their paths, and collapsing mountains. Referring to these events also as *zaiyi*, Gu Yong treated *jiu zheng* as disasters and anomalies that occur in response to identifiable conditions.[11] *Zheng*, however, can also be quite auspicious. The Many Proofs, eighth of the Nine Divisions in the "Great Plan" chapter of the *Documents*, include rain, sunshine, warmth, cold, and wind. When these occur out of order or out of balance, they are inauspicious. When they occur in sequence and in moderation, they lead to agricultural bounty, and act as emblems of the legitimacy of the ruler.[12] *Yan* overlaps with *xiao* in the processual sense of verifying the meaning of signs by pairing them with subsequent events. However, in contrast to *xiao*, *yan* on its own does not generally refer to the verifying events themselves. Titles of *tianwen* texts frequently employ *yan* in the phrase *zhanyan* 占驗, or "sign-readings and verifications."

Wuxing 五行: *Five Kinds of Action*; Five Resources (later Five Phases)[13]
Wushi 五事: Five Duties

The first sense of *wuxing* in the present work is that found in the excavated *Wuxing* (*Five Kinds of Action*) manuscript from Mawangdui. In that text, *wuxing* refer to five kinds of virtuous action: humaneness, wisdom, duty, ritual propriety, and sagacity. When all five operate together, true *de* 德 (virtue; see below) arises. Elsewhere, both *wuxing* and *wushi* are used in the sense in which they appear in the "Great Plan." *Wuxing* in the "Great Plan," the text from which the structure of the *Hanshu* "Wuxing Treatise" is derived, indicates types of materials the ruler possesses and must deliver to his subjects to maintain the prosperity of the realm as a whole, and therefore, his own power.[14] Especially after the late Western Han, the *wuxing* also appear as phases that ideally move in a fixed cycle, but may damage or disrupt (*li* 沴) one another when the cosmos or the court is out of balance. Because later texts rely on and incorporate earlier ones, it is inadvisable to systematically use one translation or the other throughout. I have for this reason generally left *wuxing* untranslated. The *wushi* are points of conduct that the ruler must maintain in ritual, and presumably at court, which is itself a ritual context. These include maintaining a reverent and proper manner in his bearing (*mao* 貌), speech (*yan* 言), vision (*shi* 視), listening (*ting* 聽), and thought (*si* 思).

Jiu 咎: Sign of blame
Fa 罰: Punishment; penalty
Ji 極: Extreme manifestation
Nie 孽: Bane
Huo 禍/祅: Calamity
Sheng 眚: Pestilence
Xiang 祥: Omen

This group of terms occurs repeatedly in a tradition (*zhuan* 傳),[15] recorded in the *Hanshu* "Wuxing Treatise." Each term, taken independently, might be rendered "omen" or "portent." In context, a dynamic relationship between the terms emerges:

> The tradition says: When in looking one does not see, this is called "not distinguishing." Its sign of blame is laxity; its penalty constant warmth; and its extreme manifestation is disease. At times there are grassy monstrosities, at times the bane of sprout-eating insects, at times misfortunes involving sheep, at times ailments of the eye, at times red pestilences or red omens. Water disrupts fire.[16]
>
> 傳曰：「視之不明，是謂不悊，厥咎舒，厥罰恆奧，厥極疾。時則有草妖，時則有蠃蟲之孽，時則有羊旤，時則有目痾，時則有赤眚赤祥。惟水沴火。」

The first three terms, sign of blame (*jiu*), penalty (*fa*), and extreme manifestation (*ji*), are a string of signs ordered by their increasing intensity. The sign of blame is the least severe. Without adequate response, the penalty comes, and if the penalty too is ignored, the extreme manifestation occurs. The sequence ends there. The entry next lists a variety of types of signs that might occur: banes (*nie*), calamities (*huo*), pestilences (*sheng*), and omens (*xiang*). These general types of signs account for the myriad varieties of specific signs found in the "Wuxing." Under the heading of lapses in the ritual Duty of Vision, the treatise includes a lack of ice in spring, the unearthing of a terracotta pot containing the remains of a creature similar to a sheep, and blood rains. Some might be present, some absent; they do not necessarily appear in a fixed order. The renderings of the terms for these types of signs in English are, admittedly, somewhat arbitrary. Calamity is a common rendering for *huo*; *nie* and *sheng* could be rendered in any number of ways so long as their inauspicious sense comes through in translation. *Xiang* elsewhere means auspicious or auspicious sign; however, in the context of the tradition cited in the "Wuxing Treatise," it is a type of inauspicious omen.

Ming 命: Charge, mandate; order, command; life, lifespan

Ming in its most basic sense refers to a command and the authorization to carry out that command. The rendering *charge* is intended to convey the sense of both power and responsibility that the term carries. Dynastic founders claimed to be in possession of *tianming* 天命, traditionally rendered as the "Mandate of Heaven," a celestial command to establish a new dynasty and the authority to rule. A secondary sense of *ming* as life or lifespan resonates with the impermanence of the charge, which may be lost due to extended periods of misrule.

De 德: Virtue, power, character; favor

De is most often translated as "virtue" based on the sense of the Latin *virtus*, a word signifying the power inherent in a given person or object, rather than the moral-ethical sense the word "virtue" has acquired in contemporary English. Arthur Waley, as early as the 1930s, suggested "charismatic power" as a more faithful translation.[17] *De* refers both to the internal moral qualities of the charismatic ruler and the external power derived from those qualities. Moreover, *de* can describe a power that is anything but moral in early imperial and pre-imperial texts, as in the phrase *xiong de* 凶德 (baleful power).[18] Where *de* is used with reference to an individual other than the ruler, the rendering "character" is sometimes employed here. This is fundamentally the same sense of *de* as suasive power but of a lesser magnitude.

De is used in a second sense in opposition to *xing* 刑 (harshness, harsh measures, or punishment) in which case it is rendered as "favor." *De* and *xing* are two basic means the ruler possesses to influence and control his subjects—the proverbial carrot and stick. The *xing-de* opposition plays a major role in administrative arguments tied to the calendar. Dong Zhongshu argues, for instance, that harshness should be employed only during times of year when *yin* permeates the cosmos so that the administration in the human realm will match and follow along with the cycles of the heavens.

I have varied my rendering of *de* based on its usage in particular passages rather than attempting to employ a single translation throughout. Where it has been workable to do so, I have retained the traditional rendering of *de* as "virtue."

Appendix B
The Bibliographic Category of *Tianwen*

An annotated translation of the titles of *tianwen* texts in the *Hanshu* bibliographic treatise is presented below.[1] While all of the texts included in the *tianwen* bibliographic category are now lost, the treatise's description of the purpose of *tianwen* texts (included in the introduction to this volume) and the titles of the texts in the imperial library produce a rough picture of what kind of materials might be labeled *tianwen* texts and a clear picture of what sort of signs they were used to read. These titles show both astronomical and meteorological aspects of the *tianwen* discourse. They include four titles referring to the Five Planets as a group, two referring to the twenty-eight lunar lodges, and others referring to particular stars or constellations, including Jade Crossbar and Grand Stairway, both in Ursa Major. Two titles refer to comets, and one to meteors. Several titles also refer to meteorological phenomena, including two concerning clouds and rain. Four titles refer to *qi*. There is no clear divide between astronomical and meteorological phenomena, and both types of phenomena are sometimes included in the title of a single text. Some titles attribute the texts to divine sources, including the Yellow God (i.e., the Yellow Emperor), Supreme Unity, and in the case of the six "Oceanic" texts, to the inhabitants of mythic isles. Yao Zhenzong's 姚振宗 (1843–1906) excellent annotated version of the treatise is an invaluable source for information on the titles.[2]

Tianwen Texts in the Bibliographic Treatise to the Hanshu

Supreme Unity and Various Masters on the Stars 泰壹雜子星, 28 j.
Five Remnants[3] and Various Stellar Aberrations 五殘雜變星, 21 j.
The Yellow God and Various Masters on Qi 黃帝雜子氣, 33 pian
Chang Cong's[4] Qi of the Sun, Moon, and Stars 常從日月星氣, 21 j.
Duke Huang and Various Masters on the Stars 皇公雜子星, 22 j.
Various Masters of Huainan on the Stars 淮南雜子星, 19 j.
Supreme Unity and Various Masters on Clouds and Rain 泰壹雜子雲雨, 34 j.
Guo Zhang Observations of Rainbows, Clouds, and Rain 國章觀霓雲雨, 34 j.
The Six Tallies of the Grand Stairway[5] 泰階六符, 1 j.
Golden Measure and Jade Crossbar: The Rise and Setting of the Five Planets, Guest Stars, and Flowing Stars in the Han[6] 金度玉衡漢五星客流出入, 8 pian
Sign-Readings and Verifications Concerning the Procession of the Five Planets, Comets, and Guest Stars in the Han 漢五星彗客行事占驗, 8 j.
Sign-Readings and Verifications Concerning the Procession of the Qi around the Sun in the Han 漢日旁氣行事占驗, 3 j.
Sign-Readings and Verifications Concerning the Procession of Meteors in the Han 漢流星行事占驗, 8 j.
Sign-Readings and Verifications Concerning the Movement of the Qi around the Sun in the Han 漢日旁氣行占驗, 13 j.
Sign-Readings and Verifications Concerning the Procession of Solar Eclipses, Lunar Halos, and Various Aberrations in the Han 漢日食月暈雜變行事占驗, 13 j.
Oceanic Account of Sign-Readings and Verifications Concerning Stars 海中星占驗, 12 j.
Oceanic Account of Various Matters Concerning the Paths[7] of the Five Planets 海中五星經雜事, 22 j.
Oceanic Account of the Prograde and Retrograde Motion of the Five Planets 海中五星順逆, 28 j.
Oceanic Account of the Domains Corresponding to the Twenty-Eight Lunar Lodges 海中二十八宿國分, 28 j.
Oceanic Account of Ministerial Roles Corresponding to the Twenty-Eight Lunar Lodges 海中二十八宿臣分, 28 j.
Oceanic Account of Various Sign-Readings Concerning the Sun, Moon, Comets, and Rainbows 海中日月彗虹雜占, 18 j.
Secret Records Concerning the Yellow River Chart and the Luo River Writings 圖書祕, 17 pian

Notes

Introduction

1. Michael Nylan, "Classics without Canonization: Learning and Authority in Qin and Han," in *Early Chinese Religion, Part One: Shang through Han (1250 BC–220 AD)*, eds. John Lagerwey and Marc Kalinowski, vol. 1 (Leiden: Brill, 2009), 726.

2. Nylan, "Classics without Canonization," 723.

3. For an essay highlighting the interrelationship between divination and textual exegesis in these two Classics, see John B. Henderson, "Divination and Confucian Exegesis," *Extrême-Orient, Extrême-Occident* 21 (1999): 79–89.

4. Liu Xiang 劉向 (79–8 BCE) led this project while serving as Superintendent of the Imperial Household (Guangluxun 光祿勳) under Emperor Cheng. Records concerning the activist editorial process he and his colleagues undertook survive in fragments of his catalogue, the *Bie lu* 別錄 (Catalogued Records) and in fragments of his son Liu Xin's 劉歆 (46 BCE–23 CE) *Qi lue* 七略 (Seven Summaries). See Michael Nylan, *Yang Xiong and the Pleasures of Reading and Classical Learning in China* (New Haven, CT: American Oriental Society, 2011), 43–47; Miranda Brown, "Liu Xiang: The Imperial Library and the Creation of the Exemplary Healer List," in *The Art of Medicine in Early China: The Ancient and Medieval Origins of a Modern Archive* (Cambridge: Cambridge University Press, 2015), 89–109. Yang Xiong 揚雄 (53 BCE–18 CE), who served under Liu Xiang in the palace library, employed the term Five Classics in his *Fayan* 法言, popularizing its use. See Fukui Shigemasa 福井重雅, *Kandai Jukyō no shiteki kenkyū: Jukyō no kangakuka o meguru teisetsu no saikentō* 漢代儒教の史的研究: 儒教の官學化をめぐる定說の再檢討 (Kyūko sōsho 60, Tokyo: Kyūko Shoin, 2005), 150–51; Nylan, "Classics without Canonization," 750–65.

5. *Wen* seems here to refer to civil methods of governance as opposed to martial methods (*wu* 武). More literally, *wen* might be rendered as "patterning," the same sense in which it is used in *tianwen* below. I read *yue wen* 說文 (delight

in civility) for *she wen* 設文 (establish civility) on analogy with *wu wen* 惡文 (detest civility) later in the text.

6. *Yanzi chunqiu* in vol. 4 of *Zhuzi jicheng* 諸子集成, 8 vols. (Beijing: Zhonghua shuju, 1952), 1.18, p. 26. NB: While the dialogue occurs in the sixth century BCE, the *Yanzi chunqiu* was compiled by Liu Xiang circa 26 BCE. The text likely represents an early imperial idealized speech of Master Yan, rather than a speech delivered by the historical figure himself. See Wu Zeyu's 吳則虞 *Yanzi chunqiu jishi* 晏子春秋集釋, vol. 1 (Beijing: Zhonghua, 1962), 49–50. A text titled *Yanzi* appears in the Classicist 儒者 subsection of the Masters 諸子 category in the *Hanshu* 漢書 (History of the Han) bibliographic treatise. See Ban Gu 班固 (32–92) et al. comp., Yan Shigu 顏師古 (581–645) comm., *Hanshu* (Beijing: Zhonghua, 1962), 30.1724.

7. Wang Xianqian 王先謙 (1842–1918) comm., *Xunzi jijie* 荀子集解, vol. 2 of *Zhuzi jicheng,* 11.17, 205.

8. Yang Bojun 楊伯峻 (1909–1992), *Chunqiu Zuozhuan zhu* 春秋左傳集注 (Beijing: Zhonghua, 1981), 4: 1390–92. For a translation of the relevant passages, see Stephen Durrant, Wai-yee Li, and David Schaberg trans., *Zuo Tradition (Zuozhuan): Commentary on the "Spring and Autumn Annals"* (Seattle: University of Washington Press, 2016), 1547–49.

9. For the human incapacity to influence Heaven, see "Aberrational Events" (Bian dong pian 變動篇) in Wang Chong, *Lun heng* 論衡 (Balanced Discourses) in vol. 7 of *Zhuzi jicheng*. For Wang's views on the effects the allotment of *qi* exercises on human beings, see his "Physiognomy of Bones" (Gu xiang 骨相), "Allotment and Fortune" (Ming lu 命祿), "Qi and Lifespan" (Qi shou 氣壽), and "Initial Endowment" (Chu bing 初稟) chapters. For a complete translation of the *Lun heng*, see Alfred Forke, *Lun-Hêng* (New York: Paragon Book Gallery, 1962).

10. Barrens corresponds to the stars β Aquarii and α Equulei. Different constellations, usually groups of lunar lodges, corresponded to different domains. One constellation corresponded to Qi, another to its neighbor Lu, another to the Kingdom of Chu, and so forth.

11. For the culture versus nature divide in modern contexts, see Bruno Latour's discussion of the "Proliferation of Hybrids" in *We Have Never Been Modern*, Catherine Porter trans. (Cambridge, MA: Harvard University Press, 1993), 1–3.

12. Jiang Xiaoyuan, *Tianxue zhenyuan* 天學真原 (Shenyang: Liaoning jiaoyu, 1991), 2.

13. Scribes in the sense used here are no mere copyists but high-level officials charged with sign-reading and record-keeping.

14. Nathan Sivin, *Granting the Seasons: The Chinese Astronomical Reform of 1280, With a Study of its Many Dimensions and a Translation of its Records*

(New York: Springer, 2009), 35. For the relations among *li*, mathematics, and constructions of the cosmos, see Christopher Cullen, "Numbers, Numeracy and the Cosmos," in *China's Early Empires: A Re-appraisal*, eds. Michael Nylan and Michael Loewe (Cambridge: Cambridge University Press, 2010), 323–38.

15. Readers familiar with celestial sciences in the Near East will no doubt notice the parity between the term *tianwen* and the Babylonian notion of "heavenly writing" (šiṭir šamê or šiṭiri šamānī). Francesca Rochberg writes that "the patterns of stars covering the sky were a celestial script." Rochberg shows, however, that the sense of this "heavenly writing" shifted over time. In the earliest contexts, it referred to "temples made beautiful 'like the stars'" and did not directly pertain to divination. In later contexts, heavenly writing "related the constellations to cuneiform signs from which one could read and derive meaning, and thus expressed the idea that written messages were encoded in celestial phenomena." See her *The Heavenly Writing: Divination, Horoscopy and Astronomy in Mesopotamian Culture* (Cambridge: Cambridge University Press, 2004), 1–2.

16. Kong Yingda 孔穎達 (574–648) et al. comp., *Zhouyi zhengyi* 周易正義, in *Shisanjing zhushu* 十三經注疏, Ruan Yuan 阮元 (1764–1849) coll., vol. 1 (Beijing: Zhonghua, 1980), 3.25c. Cf. Richard John Lynn trans., *The Classic of Changes: A New Translation of the I Ching as Interpreted by Wang Bi* (New York: Columbia University Press, 1994), 274.

17. *Zhouyi zhengyi* 7.65b.

18. I follow Harold D. Roth's assessment that the *-xun* 訓 (explication) suffix following *Huainanzi* chapter titles in the received text was likely added by the exegete Gao You 高誘 (ca. 168–212), apparently in reference to the explications of his own teacher, Lu Zhi 盧植 (d. 192 CE). See Roth's *The Textual History of the Huai-nan Tzu* (Ann Arbor: Association for Asian Studies, 1992), 43.

19. While Sima Qian is generally treated as the compiler of the *Shiji*, note that the work was initiated by his father, Sima Tan 司馬談 (d. 110 BCE).

20. For an overview of categories in the bibliographical treatise and their relationship to problems of sign-reading, see Lisa Raphals, *Divination and Prediction in Early China and Ancient Greece* (Cambridge: Cambridge University Press, 2013), 31–40.

21. *Hanshu* 30.1765.

22. Yao Zhenzong 姚振宗 (1842–1906) suspects that Guo Zhang is either the name of an observation platform (*guan*) or a person. However, there is no known individual by the name of Guo Zhang who lived during the Western Han. See Yao's *Hanshu yiwenzhi tiaoli* 漢書藝文志條理, in *Ershiwu shi yiwen jingji zhi kaobu cuibian* 二十五史藝文經籍志考補萃編, eds. Wang Chenglüe 王承略 and Liu Xinming 劉心明 (Beijing: Qinghua daxue, 2011–2012), 378. The name of the platform itself might be rendered "Shining Domain." Because the text is lost, the precise sense of its title remains unknown.

23. The term "haizhong" 海中, roughly translatable as "in the middle of the ocean," at first glance suggests that the texts were used for nautical purposes. However, "haizhong" is frequently associated with the Three Spirit Mountains (san shen shan 三神山) and the isles of the immortals in the *Shiji* and the *Hanshu*. See, e.g., Sima Qian 司馬遷 (ca. 145–ca. 86 BCE) comp., *Shiji* 史記 (Beijing: Zhonghua, 1959), 6.247, 12.455, 28.1367, 108.3086 and *Hanshu* 25A.1216, 25A.1234, 25B.1245. Moreover, the initial portion of a given title in the *Hanshu* bibliographic treatise usually states the geographic provenance of a text or attributes it to a compiler, composer, or patron. In calling these texts "haizhong," their titles suggest that they issue from the isles of the immortals. I use the rendering "oceanic" in its sense of *of or pertaining to the inhabitants of the ocean*.

24. *Hanshu* 30.1763–65.

25. *Hanshu* 26.1298; cf. *Shiji* 27.1339. A full translation may be found in David W. Pankenier, *Astrology and Cosmology in Early China: Conforming Earth to Heaven* (Cambridge: Cambridge University Press, 2013), 500.

26. Ho Peng Yoke, *The Astronomical Chapters of the Chin Shu* (Paris: Mouton & Co., 1966), 13–23, esp. 14; Daniel P. Morgan, "The Heavenly Writing," in *Monographs in Tang Official Historiography: Perspectives from the Technical Treatises of the History of Sui (Sui shu)*, eds. Daniel P. Morgan and Damien Chaussende, with the collaboration of Karine Chemla (Cham, Switzerland: Springer, 2019), 146–58.

27. Ludwig Wittgenstein, *Philosophical Investigations*, 2nd ed., G.E.M. Anscombe trans. (1958; repr. of English text with index, Oxford: Basil Blackwell, 1986), 20; Julia Kristeva, *Desire in Language: A Semiotic Approach to Literature and Art*, trans. Thomas Gora, Alice Jardine, and Leon S. Roudiez (New York: Columbia University Press, 1980), 36–41; H. Aram Veeser ed., introduction to *The New Historicism* (New York: Routledge, 1989), xi.

28. Edward H. Schafer, *Pacing the Void: T'ang Approaches to the Stars* (Berkeley: University of California Press, 1977); Lillian Lan-ying Tseng, *Picturing Heaven in Early China* (Cambridge, MA: Harvard University Press, 2011).

29. Schafer 2–3.

30. Tseng's work is informed to an extent by the earlier study and translation of the "Hymn of the Western Passage" in Tiziana Lippiello, *Auspicious Omens and Miracles in Ancient China: Han, Three Kingdoms and Six Dynasties* (Sankt Augustin: Monumenta Serica, 2001), 89–112, 249–64.

31. Pankenier's volume is a manifestly erudite work, but necessarily speculative regarding certain prehistorical developments. Pankenier dates the founding of the Xia dynasty, a period generally regarded as mythological or legendary, on the basis of the 1953 BCE conjunction. In the absence of contemporary written records, the conjunction seems to be the sole evidence for locating the Xia in time. Had it occurred fifty years later or earlier, then the founding of the dynasty would also be said to be fifty years later or earlier. Elsewhere,

Pankenier argues that the presence of archaeological sites used for observing solstices and dating circa 2100 BCE provides historical support for the "Canon of Yao" chapter of the *Documents*, in which the sage-king Yao commissions the first calendar. In my view, Sarah Allan's claim that "much of the material in the traditional historical texts is mythical in character and thus unreliable as the basis of historical reconstruction" holds, despite Pankenier's objections that Allan sometimes departs from her own guidelines. See *Astrology and Cosmology* 31, 194, 233; Sarah Allan, "Erlitou and the Formation of Chinese Civilization: Toward a New Paradigm," *Journal of Asian Studies* 66, no. 2 (May 2007): 489.

32. Marc Kalinowski, "La rhétorique oraculaire dans les chroniques anciennes de la Chine. Une étude des discours prédictifs dans le *Zuozhuan*," *Extrême-Orient, Extrême-Occident* 21 (1999): 37–65.

33. David Schaberg, *A Patterned Past: Form and Thought in Early Chinese Historiography* (Cambridge, MA: Harvard University Press, 2001), 96–97.

34. Li Wai-yee, *The Readability of the Past in Early Chinese Historiography* (Cambridge, MA: Harvard University East Asian Center, 2007), 173–202, esp. 193–95.

35. For a study focused on relations among evidence, epistemology, and argument in early China, see Garret P. S. Olberding, *Dubious Facts: The Evidence of Early Chinese Historiography* (Albany: SUNY Press, 2012).

36. Geoffrey Lloyd and Nathan Sivin, *The Way and the Word: Science and Medicine in Early China and Greece* (New Haven, CT & London: Yale University Press, 2002), 3.

37. Lloyd and Sivin, *Way and the Word*, 3.

38. Daryn Lehoux, *What Did the Romans Know? An Inquiry into Science and World Making* (Chicago: University of Chicago Press, 2012), 9.

39. Sun Xiaochun and Jacob Kistemaker, *The Chinese Sky During the Han: Constellating Stars and Society* (Leiden: Brill, 1997).

40. Christopher Cullen, *Heavenly Numbers: Astronomy and Authority in Early Imperial China* (Oxford: Oxford University Press, 2017); Daniel P. Morgan, *Astral Sciences in Early Imperial China: Observation, Sagehood and the Individual* (Cambridge: Cambridge University Press, 2017).

41. Li Ling ed., Liu Lexian 劉樂賢 assist ed., *Zhongguo fangshu gaiguan: Zhanxing juan* 中國方術概觀: 占星卷, 2 vols. (Beijing: Renmin Zhongguo, 1993); Lu Yang, *Zhongguo gudai xingzhan xue* 中國古代星占學 (Beijing: Zhongguo kexue jishu, 2008).

42. Jiang, *Tianxue zhenyuan*, 6.

43. Chen Meidong 陳美東, *Zhongguo gudai tianwenxue sixiang* 中國古代天文學思想 (Beijing: Zhongguo kexue jishu, 2008); Feng Shi 馮時, *Zhongguo tianwen kaoguxue* 中國天文考古學 (Beijing: Shehui kexue wenxian, 2001).

44. Astrophysikalisches Institut und Universitäts-Sternwarte, "Terra-Astronomy," Friedrich-Schiller-Universität Jena, www.astro.uni-jena.de/index.php/terra-astronomy.html; Ralph Neuhäuser, Dagmar L. Neuhäuser, and Thomas

Posch, "Terra-Astronomy—Understanding Historical Observations to Study Transient Phenomena," *Proceedings of the International Astronomical Union* 14, no. A30 (2018): 145–47.

45. E.g., a 2014 *Nature* article misinterpreted a common word for dust storm as "dust rain" from a comet impacting the Earth thereby producing a widely publicized but erroneous explanation for the worldwide spike in radioisotopes in the late eighth century. Other articles, often working on the basis of automated searches, have miscategorized numerous records of halo phenomena and particularly bright sunrises or sunsets as aurorae. Images of manuscripts concerned with a broad range of celestial phenomena are routinely cropped to appear as if they contained only stars or comets. For the erroneous comet explanation, see Liu et al., "Mysterious Abrupt Carbon-14 Increase in Coral Contributed by a Comet," *Nature* 4: 3728 (Jan. 2014): 1–4, and the critique in J. Chapman, M. Csikszentmihalyi, and R. Neuhäuser, "The Chinese Comet Observation in AD 773 January," *Astronomische Nachrichten* 335, no. 9 (Nov. 2014): 964–67; for an overview of the miscategorization problem, see D. L. Neuhäuser, R. Neuhäuser, and J. Chapman, "New Sunspots and Aurorae in the Historical Chinese Text Corpus?: Comments on Uncritical Digital Search Applications," *Astronomische Nachrichten* 339, no. 1 (Jan. 2018): 10–29; for the image-cropping problem, see Jesse J. Chapman, "Ambiguity, Scope, and Significance: Difficulties in Interpreting Celestial Phenomena in Chinese Records," *Proceedings of the International Astronomical Union* 14, no. A30 (2018): 152–55.

46. This book touches at times on pre-imperial materials, insofar as they remained influential in the Western and Eastern Han. It does not, however, engage at all with the rich body of divinatory texts dating to the Shang period (ca. 1600–1046 BCE), the oracle-bone inscriptions. For a concise introduction to those materials, see Robert Eno, "Deities and Ancestors in Early Oracle Inscriptions," in *Religions of China in Practice*, ed. Donald S. Lopez, Jr. (Princeton, NJ: Princeton University Press, 1996), 41–51. For a foundational survey, see David N. Keightley, *Sources of Shang History: The Oracle-Bone Inscriptions of Bronze Age China*, 2nd ed. (Berkeley: University of California Press, 1985).

47. For a recent and concise description of diviner's boards, see Marc Kalinowski, "The Notion of 'shi' 式 and Some Related Terms in Qin-Han Calendrical Astrology," *Early China* 35–36 (2012–13): 331–60. A number of excellent new studies of hemerological texts are included in Donald Harper and Marc Kalinowski eds., *Books of Fate and Popular Culture in Early China: The Daybook Manuscripts of the Warring States, Qin, and Han*, Handbuch der Orientalistik IV.33 (Leiden & Boston: Brill, 2017). For the lunar lodges as a day count, see Marc Kalinowski, "The Use of the Twenty-Eight *xiu* as a Day-Count in Early China," *Chinese Science* 13 (1996): 55–81.

48. Howard L. Goodman, "Chinese Polymaths, 100–300 AD: The Tung-Kuan, Taoist Dissent, and Technical Skills," *Asia Major* 3rd ser., 18.1 (2005): 164.

49. Part II places greater emphasis on sources for the Eastern Han, especially Fan Ye 范曄 (398–445) comp., Li Xian 李賢 (651–684) comm., *Hou Han shu* 後漢書 (Beijing: Zhonghua, 1965). *Hou Han shu* technical treatises (*zhi* 志) are compiled by Sima Biao 司馬彪 (240–306), with commentary by Liu Shao 劉昭 (fl. ca. 502–ca. 519).

Chapter 1

1. Despite the general rule that only members of the Liu clan could serve as kings, Wu Rui 吳芮 (d. 201) and his descendants maintained their kingdom in Changsha for five generations, until 157. Wu Rui was one of seven kings who had encouraged Liu Bang to take the title of emperor in 202. See Michael Loewe, *A Biographic Dictionary of the Qin, Former Han, and Xin Periods: 221 BC–24 AD* (Leiden: Brill, 2000), 586; *Hanshu* 34.1894.

2. *Shiji* 19.978. Li Cang 利蒼 is the name found on a seal in the second tomb at Mawangdui. In the *Shiji*, his personal name is written Cang 倉, and in the *Hanshu*, he is identified as Li Zhucang 黎朱蒼 (16.618). The first tomb contains the remains of his wife, Xin Zhui 辛追, identified on the basis of a wooden seal. For a summary of what is known concerning the three tomb occupants, see Luke Waring, "Writing and Materiality in the Three Han Dynasty Tombs at Mawangdui" (PhD diss., Princeton University, 2019), UMI (13864026), 27–31.

3. For an overview of the decades leading up to the Qin conquest, see Derk Bodde, "The State and Empire of Ch'in," in *The Ch'in and Han Empires, 221 B.C.–A.D. 220*, eds. Dennis Twitchett and Michael Loewe, vol. 1 of *The Cambridge History of China* (Cambridge: Cambridge University Press, 1986), 40–45. For the fall of the Qin, see ibid. 81–85.

4. For Liu Bang's establishment of the Western Han, see Michael Loewe, "The Former Han Dynasty," in Twitchett and Loewe, *Ch'in and Han Empires*, 119–23. For Liu Bang's death, see Loewe, *Biographical Dictionary*, 258; *Hanshu* 1B.78–79.

5. Empress Lü 呂 (d. 180 BCE) effectively ruled the empire for a fifteen-year period following Liu Bang's death in 195. During her reign, she transferred considerable power to her own clan, appointing two of her brothers to the two highest official positions, Chancellor of State (*xiang guo* 相國) and Supreme General (*shang jiangjun* 上將軍). After her death, they launched a failed attempt to remove the Liu clan from power (Loewe, "Former Han Dynasty," 136).

6. A bamboo slip in the tomb indicates that its occupant was interred in 168 BCE, the twelfth year of Emperor Wen 文 (r. 180–157). Archaeologists have generally agreed that the occupants of tombs one and two, Xinzhui and Li Cang, were the parents of the occupant of tomb three. Because early historians Sima Qian and Ban Gu both give the first year of the third Marquis, Pengzu 彭

祖, as 164 BCE, there is some doubt as to whether or not the second Marquis should be identified as the tomb occupant (see *Shiji* 19.978 and *Hanshu* 16.618). He Jiejun 何介鈞 notes that the tomb occupant had only a three-layer coffin, rather than the seven-layer coffin befitting a Marquis (*hou* 侯) or Marquise. See his *Changsha Mawangdui er san hao Hanmu* 長沙馬王堆二三號漢墓 (Beijing: Wenwu, 2004), 237–40. Fu Juyou 傅舉有 argues that the abundance and high quality of the goods in the tomb indicate that its occupant was none other than the second Marquis. Fu suspects that the date given for his death in received sources is an error. See "Han dai liehou de jiali—jian tan Mawangdui sanhao mu muzhu" 漢代列侯的家吏——兼談馬王堆三號墓墓主, *Wenwu* 文物 1 (1999): 96. Li Shisheng 黎石生 has argued that, in the absence of compelling evidence to the contrary, the theory that the tomb occupant was a younger brother of Li Xi is most likely to be true. See "Changsha Mawangdui sanhao muzhu zaiyi" 長沙馬王堆三號墓主再議, *Gugong bowuyuan yuankan* 故宮博物院院刊, no. 3 (2005): 150–55, 162. Chen Songchang 陳松長, in collaboration with the seal expert Sun Weizu 孫慰祖, has reconstructed incomplete characters on a damaged seal as Li Xi 利豨. See Waring, "Writing and Materiality," 30 and Chen Songchang, "Mawangdui sanhao muzhu de zai renshi" 馬王堆三號墓主的再認識, in *Jianbo yanjiu wengao* 簡帛研究文稿 (Beijing: Xianzhuang shuju, 2008), 435–42. Regardless of the precise identity of the tomb occupant, we can infer from his surroundings, and from the kinship relations of the occupants of tombs one and two, that he was a high status member of the Li clan and likely served at the Changsha court.

7. *Shiji* 19.978–79; Loewe, *Biographical Dictionary*, 219.

8. Recent scholarship has cast doubt on claims that tomb texts were routinely produced as funerary objects (*mingqi* 明器). See Waring's concise summary of scholarship on reasons for interring texts in Han tombs, including the possibility that some were produced specifically as funerary objects, "Writing and Materiality," 447–48; for a refutation of tomb texts as *mingqi*, see Alain Thote, "Daybooks in Archaeological Context," in *Books of Fate and Popular Culture in Early China: The Daybook Manuscripts of the Warring States, Qin, and Han*, eds. Donald J. Harper and Marc Kalinowski (Leiden: Brill, 2017), 40–47.

9. For annotated translations of these texts, see Donald Harper, *Early Chinese Medical Literature: The Mawangdui Medical Manuscripts* (London: Kegan Paul, 1998).

10. Sign-reading was a critical skill for commanders as they are described in military texts from early China. For sign-reading in early military contexts, see Albert Galvany, "Signs, Clues and Traces: Anticipation in Ancient Chinese Political and Military Texts," *Early China* 38 (2015): 1–43.

11. I draw on Matthias L. Richter's discussion of reading and transmission practices surrounding "Min zhi fu mu" 民之父母 (Father and Mother to the People) in *The Embodied Text: Establishing Textual Identity in Early Chinese Manuscripts* (Leiden: Brill 2013), 189–190.

12. Richter, *Embodied Text*, 188.

13. Cf. Richter, *Embodied Text*, 192.

14. Both detailed transcriptions and excellent images are included in Qiu Xigui 裘錫圭 ed., *Changsha Mawangdui Hanmu jianbo jicheng* 長沙馬王堆漢墓簡帛集成, 7 vols. (Changsha; Beijing: Zhonghua shuju, 2014).

15. Xu Zhentao made major contributions to the use of historical records from China to study astronomical phenomena, and his work in English is frequently cited by scholars beyond the early China field. More than 120 publications, primarily in the fields of astrophysics, terra-astronomy, and cultural astronomy, have cited Xu Zhentao, David W. Pankenier, and Jiang Yaotiao, *East Asian Archaeoastronomy: Historical Records of Astronomical Observations of China, Japan, and Korea* (Amsterdam: Gordon and Breach Science Publishers on behalf of the Earth Space Institute, 2000).

16. Xu Zhentao, "Cong boshu 'Wuxing zhan' kan 'Xian-Qin hunyi' de chuangzhi" 從帛書《五星占》看"先秦渾儀"的創制, *Kaogu* 考古 (Feb. 1976): 84, 89–94; Xi Zezong, "Mawangdui Hanmu boshu zhong de huixingtu" 馬王堆漢墓帛書中的彗星圖, *Wenwu* 文物 2 (1978): 5–9; Gu Tiefu, "Mawangdui boshu 'Tianwen qixiang zazhan' neirong jianshu" 馬王堆帛書天文氣象雜占內容簡述, *Wenwu* 2 (1978): 1–4.

17. Early usage of the term "textual community" focused on monastic communities in medieval Europe and the relationships among textual transmission, copying, reading, and orality in that context. See, e.g., Brian Stock, *The Implications of Literacy: Written Language and Models of Interpretation in the Eleventh and Twelfth Centuries* (Princeton, NJ: Princeton University Press, 1983); Duncan Robertson, "Writing in the Textual Community: Clemence of Barking's Life of St. Catherine," *French Forum* 21, no. 1 (Jan. 1996): 5–28. The present work, however, uses the term in a more capacious manner, allowing for any community that shares a given corpus of works to be defined as a textual community. In this regard, I follow Kuisma Korhonen, "Textual Communities: Nancy, Blanchot, Derrida," *Culture Machine* 6 (2006). http://culturemachine.net/community/textual-communities

18. Waring suggests that the Mawangdui tombs were likely located in Linxiang 臨湘, an eastern suburb of the Changsha capital, but notes some uncertainty concerning the precise location of that city. See "Writing and Materiality," 40–41, including note 57.

19. For a concise English language summary of the contents and dating of tomb 2 and tomb 3, see He Jiejun, *Changsha*, 382–87.

20. Transcriptions here follow those of Liu Jianmin 劉建民, "Wuxing zhan" 五星占, in Qiu, *Changsha Mawangdui*, 4: 223–44. I have also consulted transcriptions in Liu Lexian 劉樂賢, "'Wuxing zhan' kaoshi" 五星占考釋 in *Mawangdui tianwen shu kaoshi* 馬王堆天文書考釋 (Guangzhou: Zhongshan daxue, 2004), 29–99 and the earlier report collectively authored by the Mawangdui

archaeological team, "Mawangdui Hanmu boshu 'Wuxing zhan' shiwen" 馬王堆漢墓帛書〈五星占〉釋文, *Zhongguo tianwenxue shi wenji* 中國天文學史文集 (Beijing: Kexue, 1979), 1–13.

21. See Christopher Cullen, "*Wu Xing Zhan* 五星占 'Prognostics of the Five Planets,'" *SCIAMVS: Sources and Commentaries in the Exact Sciences* 12 (2011): 193–249; Christopher Cullen, "Understanding the Planets in Ancient China: Prediction and Divination in the *Wu xing zhan*," *Early Science and Medicine* 16 (2001): 218–51; and Daniel Patrick Morgan, "The Planetary Visibility Tables in the Second-Century BC Manuscript Wu Xing Zhan 五星占," *East Asian Science, Technology, and Medicine* 43 (2016): 17–60. For an annotated translation of the manuscript into modern Chinese, see Chen Jiujin 陳久金, *Boshu ji gudian tianwen shiliao zhu xi yu yanjiu* 帛書及古典天文史料注析與研究 (Taipei: Wanjuan lou 2001), 102–47.

22. There are three separate manuscripts from Mawangdui that archaeologists have labeled "Xingde." The third is in quite poor condition. Transcriptions of the more complete two are drawn from Cheng Shaoxuan, "Xingde jia" in Qiu, *Changsha Mawangdui*, 5: 1–30 and "Xingde yi" in Qiu, *Changsha Mawangdui*, 5: 31–48. Each manuscript contains two major sections: a meteoromantic section referred to as the "Ri yue feng yu yunqi zhan" 日月風雨雲氣占 (Readings of the Sun, Moon, Wind, Rain, and Cloud *Qi*) and a Nine Palaces diagram with mantic passages pointing to the significance of events occurring on days associated with Xing (Harshness), De (Favor), and various deities, referred to as the "Xingde zhan" 刑德占 (Harshness and Favor Sign-Readings). For a brief summary of the debate regarding whether the two sections in fact constitute separate texts, see Cheng Shaoxuan, "Xingde jia," 1. Whether the manuscripts each contain one text or two, I will consider them as whole objects, as the contents of the sections are related and they were available to the same people at the same time. I have also consulted transcriptions in Chen Songchang, *Mawangdui boshu "Xingde" yanjiu lungao* 馬王堆帛書《刑德》研究論稿 (Taibei: Taiwan guji, 2001) and, for the meteoromantic section, Liu Lexian, "Ri yue feng yu yunqi zhan," in *Mawangdui tianwen*, 160–94. For the technical intricacies of the Nine Palaces diagram and "Xingde" portion of the manuscript, see Marc Kalinowski, "The Xingde 刑德 Texts from Mawangdui," trans. Phyllis Brooks, *Early China* 23–24 (1998–99): 125–202.

23. Kalinowski, "*Xingde* Texts," 130.

24. Liu Lexian employs this title and does not discuss other parts of the "Harshness and Favor" manuscripts in his volume on Mawangdui *tianwen* texts. Chen Songchang notes that several titles have been applied to the "Meteoromantic Sign-Readings," including "Xingzhan shu" 星占書 (Writing on Sign-Readings for the Planets and Stars); "Jun zazhan" 軍雜占 (Miscellaneous Sign-Readings on Military Matters); and "Tianxiang zazhan" 天象雜占 (Miscellaneous Sign-Readings on Celestial Images) in addition to his preferred "Yunqi zhan" 雲氣占 (Sign-Readings on Cloud *Qi*) (*Mawangdui boshu "Xingde"* 58).

Notes to Chapter 1 | 197

25. Transcriptions are drawn from Wang Shujin 王樹金, "Tianwen qixiang zazhan" 天文氣象雜占, in Qiu, *Changsha Mawangdui*, 4: 245–75. Rank and entry number are provided following the page number, e.g., 246, line 1.2, refers to the second entry in the first rank on p. 246. I have also consulted Liu Lexian's annotated transcription, "Tianwen qixiang zazhan," in *Mawangdui tianwen*, 100–159 and Yamada Keiji's 山田慶兒 transcription and translation, "Tenmon kishō zatsusen" 天文氣象雜占 in *Shinhatsugen Chūgoku kagakushi shiryō no kenkyū* 新發現中國科學史資料の研究 (Kyōto: Kyōto daigaku jinbun kagaku kenkyūjo, 1985), 45–86.

26. See, e.g., Chen Meidong, *Zhongguo gudai tianwenxue sixiang*, 388 and images presented in Feng Shi, *Zhongguo tianwen*, 254–55.

27. As Donald Harper has noted, users of the "Miscellaneous Sign-Readings" could match sign-reading statements in this section to images in the main body of the manuscript. See "Communication by Design: Two Silk Manuscripts of Diagrams (*Tu*) from Mawangdui Tomb Three," in *Graphics and Text in the Production of Technical Knowledge in China: The Warp and the Weft*, eds. Francesca Bray, Vera Dorofeeva-Lichtmann, and Georges Métailié (Leiden: Brill, 2007), 174.

28. For a summary of early scholarship on this manuscript, see Cullen, "Understanding the Planets," 222–23.

29. Liu Jianmin, "Wuxing zhan," 231, line 39a.

30. It is likely that the "Five Planets Sign-Readings" was compiled from multiple texts. The order of presentation of the planets indicates the composite nature of the text. Like later treatises, the "Five Planets" associates days 7 and 8 (*geng* 庚 and *xin* 辛) with Venus and days 9 and 10 (*ren* 壬 and *gui* 癸) with Mercury. However, Mercury appears *before* Venus in the manuscript. The inverted order, paired with the correspondences to particular days, suggests that a copyist working with a damaged manuscript reversed the order of the sections, ignoring or failing to notice the corresponding days. Odd as this order may seem, images of the manuscript incontrovertibly show the transition from the end of the section on Saturn directly to the beginning of the section on Mercury (Qiu, *Changsha Mawangdui*, 1: 173). Moreover, the second presentation of correspondences to directions is redundant within the context of the "Five Planets," though it probably was not in source texts. Finally, the manuscript uses multiple names for certain planets (discussed later in the main text for this chapter).

31. Reading *shi* 失 as *die* 昳 following Liu Lexian ("Wuxing zhan," 86, n.1).

32. Liu Jianmin, "Wuxing zhan," 236, lines 74b–75a. However, I follow Liu Lexian ("Wuxing zhan," 86, line 76) in transcribing *si tian xian* 司天獻 (Master of Celestial Sacrifices) where Liu Jianmin has *si shi xian* 司失獻 (Master over Faults in Sacrifice). The same pattern holds true for the taxonomic associations paired with each planet.

33. The phrase "the center is allotted to Soil" (*zhongyang fen tu* 中央分土) occupies the same position within the section on Saturn where seasonal

correspondences are given for other planets (Liu Jianmin 230, line 31a; Cullen 228, #52).

34. Liu Jianmin, "Wuxing zhan," 224, lines 7b–8a; cf. Cullen, "Five Planets," 222, #21. (Titles are omitted in further references to this pair of sources.)

35. Liu Jianmin 230, lines 30b–31a; cf. Cullen 228, #51.

36. Liu Jianmin 230, line 30a; cf. Cullen 228, #50.

37. Liu Jianmin 230, line 29b; cf. Cullen 227, #47.

38. Liu Jianmin 224, lines 8b–9a; cf. Cullen 223, #23.

39. See, e.g., memorials discussing the earthquake of 133 CE in chapter 5, pp. 160–70.

40. Liu Jianmin 231, line 33; cf. Cullen 229, #58.

41. Liu Jianmin 230–31, lines 32b–33a; cf. Cullen 229, #55.

42. Liu Jianmin 230, line 32b; cf. Cullen 228, #54.

43. Liu Jianmin 232, lines 42a–42b; cf. Cullen 232, #71.

44. Liu Jianmin 231, line 33b; Cullen 229, #57. Both Chamber and Heart correspond to stars in Scorpius.

45. In the manuscript, the character *hui* 槶 appears with the *mu* 木 element on the lower portion of the character, rather than on its left side.

46. Liu Jianmin 231, line 39a; cf. Cullen 230, #67.

47. Liu Jianmin 229, line 24a; cf. Cullen 226, #37.

48. Liu Jianmin 229, lines 24b–25a; cf. Cullen 226, #39.

49. Liu Jianmin 229, line 24a; cf. Cullen 226, #38.

50. Liu Jianmin 229, line 23b; cf. Cullen 226, #36.

51. For an additional example of a contemporary excavated technical text evidently compiled from multiple sources, see the discussion of the Kongjiapo 孔家坡 day-book in Ethan Harkness, "The Elements of the Agricultural Year," in "Cosmology and the Quotidian: Day Books in Early China" (PhD diss., University of Chicago, 2011), UMI (3487616), 161–200.

52. Liu Jianmin 226, lines 12a and 14a; cf. Cullen 224, #26 and #28.

53. Liu Jianmin 227, lines 17b–18b; cf. Cullen 224–25, #31.

54. Liu Jianmin 233, lines 61a–62b and 235, lines 63b–64b; cf. Cullen 237, #104–6, #108.

55. Liu Jianmin 230, line 32a and 231, line 35a; cf. Cullen 228, #53, 230, #62.

56. Liu Jianmin 232, lines 48a–51a; cf. Cullen 233–34, #80–#85.

57. Morgan, "Planetary Visibility Tables"; Cullen, "Understanding the Planets."

58. Cullen, "Understanding the Planets," 238–43.

59. Cullen, "Understanding the Planets," 248–49. Cullen cites David Brown, *Mesopotamian Planetary Astronomy-Astrology* (Groningen: Brill, 2000), 146–56.

60. Wang Shujin, "Tianwen qixiang zazhan," 246, line 1.1.

61. Wang Shujin, "Tianwen qixiang zazhan," 262, line 4.14. In the *Zuozhuan*, a series of signs appearing between 512 and early 510 encouraged the Wu leadership to invade Chu. These included internal divisions among the Chu leadership, a song heard in a dream, and a solar eclipse. See *Chunqiu Zuozhuan zhu*, 4: 1509, 1513–14; Durrant, Li, and Schaberg, *Zuo Tradition*, 1709–11, 1717. Given the discrepancy between the solar eclipse and the halo around the sun, there is some doubt as to whether the manuscript referred to the 506 BCE attack. Wang notes that, in any case, the omenological tradition referred to in the *Zuozhuan* differed, at least in its technical features, from that in the "Miscellaneous Sign-Readings" manuscript (264, n.9).

62. Wang Shujin, "Tianwen qixiang zazhan," 254, lines 2.42–2.45.

63. Wang Shujin, "Tianwen qixiang zazhan," 264, line 4.57. The last character *sa* 卅 (thirty) might also be read as *ce* 冊 (strips). Thus, the full sentence would read, "The world loses its master and the strips are divided."

64. Wang Shujin, "Tianwen qixiang zazhan," 270, line 6.12.

65. Wang Shujin, "Tianwen qixiang zazhan," 262, line 4.13.

66. Wang Shujin, "Tianwen qixiang zazhan," 269, lines 6.1–6.2.

67. Wang Shujin, "Tianwen qixiang zazhan," 271, line 6.41.

68. Wang Shujin, "Tianwen qixiang zazhan," 270, line 6.29. The punctuation mark = doubles *shi* 是, here and in line 6.25 below.

69. Wang Shujin, "Tianwen qixiang zazhan," 270, line 6.25.

70. Wang Shujin, "Tianwen qixiang zazhan," 261, lines 4.1–4.4, slightly modified, following Liu Lexian's interpretation of the semantic meaning of the character *bo* 柏. See Liu's "Tianwen qixiang," 119. Wang Shujin interprets *bo* to mean *ba* 霸 (hegemon).

71. "When troops are in the field, should there be lightning on sexagenary day 1, then in no more than one hundred days the troops will engage in battle" 兵在埜 (野), 甲子䨻 (雷 〈雨〉), 不出百日兵入 (Cheng Shaoxuan, "Xingde jia," 15, line 43). I do not follow Cheng's reading of *lei* 䨻 (lightning) as referring to rain (*yu* 雨). I read *ru* 入 as "entering into battle." Liu Lexian suspects it means to re-enter the troops' home city ("Ri yue feng yu," 187, lines 43–44).

72. Cheng Shaoxuan, "Xingde jia," 16, lines 48–49.

73. The sexagenary days associated with them are as follows: Great Sound: 4, 10, 16, 22, 34, 40, 46, 52; Fenglong: 2, 8, 14, 20, 32, 38, 44, 50; Lord Thunder: 5, 11, 17, 23, 35, 41, 47, 53; Master of Rain: 6, 12, 18, 24, 36, 42, 48, 54; Earl of Wind: 3, 9, 15, 21, 33, 39, 45, 51.

74. E.g., "Far Roaming" ("Yuan you" 遠遊) features Fenglong, Earl of Wind, Lord Thunder, and Master of Rain, with Fenglong also appearing in both "Li sao" and "Longing for the Beautiful One" ("Si mei ren" 思美人). See Hong Xingzu 洪興祖 (1090–1155) and Wang Yi 王逸 (fl. 89–158) comms., *Chuci buzhu* 楚辭補注 (Beijing: Zhonghua, 1983), 5.168, 170–1, 1.31, 4.147. Fenglong also appears

in poems by Sima Xiangru 司馬相如 (179–117) (*Hanshu* 57B.2598), Yang Xiong (*Hanshu* 87A.3519, 3538), and Zhang Heng (*Hou Han shu* 59.1933). Both Earl of Wind and Master of Rain appear in the *Shiji* "Fengshan Treatise" ("Fengshan shu" 封禪書; 28.1375), and, along with Lord Thunder, in the *Hanshu* "Treatise on the Suburban Sacrifices" ("Jiao si zhi" 郊祀志; 25A.1206–7, 25B.1268).

75. Cheng Shaoxuan, "Xingde jia" 29, lines 136–37.

76. Cheng Shaoxuan, "Xingde jia" 2, line 8. The manuscript punctuation mark ⹀ doubles the character jun 軍, with the first understood as meaning yun 暈 (halo).

77. Cheng Shaoxuan, "Xingde jia," 3, line 13.

78. Cheng Shaoxuan, "Xingde jia," 14, line 38. Liu Lexian omits the *xin* 心 ("Ri yue feng yu," 185, line 38), whereas Chen Songchang includes it (*Xing-De yanjiu*, 152). Cheng Shaoxuan's transcription follows Chen's reading of the line (as does my own). However, this reading remains tentative, and I have included a question mark in both the transcription and translation. Examining the manuscript, while the character *xin* is clear, a single curved downward line seems to indicate the character *ta* 它, meaning "other." Moreover, there is no separation between this line and the character Chen reads as *xin* above it; Liu's decision to read a single graph here may well be correct. However, in my judgment, it better resembles a *xin* graph than *ta*. For the image, see Qiu, *Changsha Mawangdui*, 1: 212, line 38. "Xingde yi" unfortunately cannot be used to arbitrate between the two transcriptions as the relevant portion is missing (Cheng Shaoxuan 46, line 86).

79. Following Liu Lexian ("Ri yue feng yu," 186, n.3), Cheng Shaoxuan suggests that *ri* 日 (sun) is a mistranscription for *bai* 白 (white), placing *bai* in triangular brackets <白> following the character *ri*. Chen Songchang simply transcribes the character as *bai* 白 (*Xing-De yanjiu*, 153). The manuscript has *ri*, however, and I have transcribed and translated the text as it stands.

80. Cheng Shaoxuan, "Xingde jia," 14, lines 38–39. Chen Songchang separates *nai ju* 乃矍 and *baixuan* 百軒 with a comma and omits the period before *sanjian* 三見 (*Xing-De yanjiu*, 152). Following his reading, the last few phrases might be rendered: ". . . then in that direction there will be fear. If it should appear three times before the hundred houses, then blood will flow and bones will bleach in the sun."

81. As Stephan N. Kory has noted, the common term *zhan hou* 占侯 (perhaps best rendered by Kalinowski as meteoromancy) for reading signs in atmospheric *qi* did not come into use until the Eastern Han. However, passages such as this attest to the practice of reading these types of signs far earlier. For a systematic study of the term *zhan hou* and its various uses in Eastern Han and later sources, see Kory's "Omen Watching, Mantic Observation, Aeromancy, and Learning to 'See,'" *East Asian Science, Technology, and Medicine* 50 (2019): 67–132.

82. Transcriptions of the *Constant Models* are drawn from Hirose Kunio 廣瀬薰雄, *Jing fa* 經法 (Constant Models), in Qiu, *Changsha Mawangdui*, 4: 125–50. Titles of subsections are provided in individual notes. A complete translation is presented under the title "The Canon: Law" in Robin D. S. Yates, *Five Lost Classics: Tao, Huang-Lao, and Yin-yang in Han China* (New York: Ballantine, 1997), 49–101.

83. The *Daodejing* is commonly read outside China, even by non-specialists, so I will not review its contents here. For unfamiliar readers, I recommend Arthur Waley's translation, *The Way and Its Power: A Study of the Tao Te Ching and Its Place in Chinese Thought* (New York: Grove, 1958 [1934]). For the Mawangdui versions of the *Daodejing*, see Robert G. Henricks trans., *Lao-tzu: Te-tao ching* (New York: Ballantine, 1989). Note that the two chapters of the text appear in reverse order in the Mawangdui versions. The *Constant Models* is appended to *Laozi* B, along with another text titled *Sixteen Constants* (Shiliu jing 十六經).

84. Jeffrey L. Richey, "Lost and Found Theories of Law in Early China," *Journal of the Economic and Social History of the Orient* 49, no. 3 (2006): 336. NB: While Richey's article is generally helpful, the contents of the Mawangdui tombs do not support his division between "Huang-Lao" and Confucian thought. Rather, the tombs show that texts associated with both types of thought circulated in the same communities, even prior to the supposed "synthesis of traditions" (343) effected by Dong Zhongshu.

85. Michael Puett, "Sages, Ministers, and Rebels: Narratives from Early China Concerning the Initial Creation of the State," *Harvard Journal of Asiatic Studies* 58, no. 2 (Dec., 1998): 455–58.

86. Hirose, "Lun" 論 (Assessment), *Jing fa*, 140, line 49a; cf. Yates, *Five Lost Classics*, 81.

87. Hirose, "Sidu" 四度 (Four Measures), *Jingfa*, 138, lines 43b–44a; cf. Yates, *Five Lost Classics*, 77. The manuscript uses a rare variant for *ji* 稽.

88. Hirose, "Daofa" 道法 (Models of the Way), *Jingfa*, 127, lines 6a–6b; cf. Yates, *Five Lost Classics*, 53.

89. Hirose, "Jun zheng" 君正 (The Lord Rectifies), *Jingfa*, 132, 19a–19b; cf. Yates, *Five Lost Classics*, 63.

90. Hirose, "Wang Lun" 亡論 (Assessing Destruction), *Jingfa*, 143, line 57b; cf. Yates, *Five Lost Classics*, 89.

91. Hirose, "Wang Lun," *Jingfa*, 143, lines 60a–60b; cf. Yates, *Five Lost Classics*, 89.

92. Hirose, "Sidu," *Jingfa*, 138; cf. Yates, *Five Lost Classics*, 77.

93. Robin D. S. Yates, "The History of Military Divination in China," *East Asian Science, Technology, and Medicine* 24 (2005): 17.

94. The version of the "Appended Statements" found at Mawangdui does not have a title and differs substantially from the received text. Much of

the material now found in the second *juan* of the received text corresponds to passages in other *Changes*-related texts found at Mawangdui.

95. For the *Changes* and related manuscripts, I generally follow transcriptions in Chen Jian 陳劍, *Zhouyi jingzhuan* 周易經傳 (Changes of Zhou: Classic and Traditions), in Qiu, *Changsha Mawangdui*, 3: 3–162. For "Xici" and "Er san zi wen," I have also consulted Zhang Zhenglang's 張政烺 handwritten notes, posthumously published as *Mawangdui boshu Zhouyi jingzhuan jiaodu* 馬王堆帛書周易經傳校讀 (Beijing: Zhonghua, 2008) and for "Yao," I have consulted Chen Songchang 陳松長 and Liao Mingchun 廖名春, "'Yao' shi wen" 《要》釋文, *Daojia wenhua* 道家文化 3 (1993): 434–35. I have also employed and in some instances followed translations in Edward L. Shaughnessy, *I Ching, The Classic of Changes: The First English Translation of the Newly Discovered Second-Century B.C. Mawangdui Texts* (New York: Ballantine, 1997). The titles "Muhe" and "Yao" are found in the texts themselves, but "Xici" and "Er san zi wen" are conventional titles applied by modern scholars. For an English language overview of these materials, see Shaughnessy, "The Mawangdui *Yijing* Manuscripts," in *I Ching*, 14–29.

96. These two hexagrams are better known by their names in the received *Changes*, respectively Qian 乾 and Kun 坤. Zhang Zhenglang's 張政烺 marginal emendations suggest that he views these as orthographic variants (*Mawangdui boshu Zhouyi*, 110).

97. Chen Jian, "Xici" 繫辭, *Zhouyi jingzhuan*, 60, line 1a.

98. Chen Jian, "Xici," *Zhouyi jingzhuan*, 61, line 3a.

99. As Willard Peterson has argued, the "Appended Statements" provides a theoretical basis for claims that the mantic technique of the *Changes* "duplicates relationships and processes at work in the realm of heaven-and-earth" and that "these relationships and processes are knowable, and the knowledge is the basis for efficacious action in the realm of human society." The *Changes* thus represents a way of understanding both the natural and human worlds that exceeds all others: "Because the knowing available to us from the *Change* resembles heaven-and-earth, and just as heaven-and-earth includes all that is, the *Change* is the guide to which all other ways are subordinate." See his "Making Connections: 'Commentary on the Attached Verbalizations' of the *Book of Change*," *Harvard Journal of Asiatic Studies* 42, no. 1 (Jun. 1982): 85, 101.

100. Zhang Zhenglang has *yuan* 原 where Chen Jian has *guan* 觀 (*Mawangdui boshu Zhouyi*, 110).

101. Chen Jian, "Xici," *Zhouyi jingzhuan*, 63, lines 6a–6b.

102. The text has a very rare character consisting in *you* 酉 + *yang* 昜, for which I have not located a Unicode listing. I follow Chen in reading this as a rare variant of *shang* 觴.

103. Chen Jian, "Er san zi wen," *Zhouyi jingzhuan*, 45, lines 12b–13a. Zhang Zhenglang has *xin* 心 where Chen Jian has *cong* 悤, parenthetically glossed as *zong* 總 (*Mawangdui boshu Zhouyi*, 95–96). The character in the manuscript image

appears closer to *xin* in my judgment, and I have accordingly followed Zhang's earlier transcription. For the image, see Qiu, *Changsha Mawangdui*, 1: 15, line 13a.

104. Chen Jian, "Er san zi wen," *Zhouyi jingzhuan*, 47, lines 18a and 18b. I do not follow Chen's parenthetical gloss of *zen* 譖 as *jian* 漸 (gradual). While this character is relatively clear on the silk manuscript, Zhang Zhenglang nonetheless pencils in a question mark next to it, perhaps due to the odd word usage (*Mawangdui boshu Zhouyi*, 97–8). I follow Shaughnessy's understanding of the word *zen* 譖 as "warning" (*I Ching*, 177).

105. Chen Jian, "Muhe" 繆和, *Zhouyi jingzhuan*, 122, line 1a, omitting Chen's parenthetical gloss of *jie* 階 as *ji* 机 here and below. I have consulted and borrowed from Richard John Lynn's translation of the hexagram Dispersion in the received text. See his *The Classic of Changes*, 513.

106. Chen Jian, "Muhe," *Zhouyi jingzhuan*, 122, lines 2a and 4a.

107. Chen Jian, "Yao," *Zhouyi jingzhuan*, 116, lines 12b–13b.

108. Earlier transcriptions of this passage have *de* 德 for *ci* 辭. Chen Jian, "Yao," *Zhouyi jingzhuan*, 117, n.11.

109. Chen Jian, "Yao," *Zhouyi jingzhuan*, 116, lines 14a. I follow Chen and Liao in parenthetically glossing *yu* 於 as *e* 閼 ("'Yao' shiwen," 434).

110. Chen Jian, "Yao," *Zhouyi jingzhuan*, 116, lines 15b–16a.

111. *Shiji* 4.119.

112. Chen Jian, "Yao," *Zhouyi jingzhuan*, 119, line 21b.

113. Chen Jian, "Yao," *Zhouyi jingzhuan*, 119, lines 22a–23b.

114. Chen Jian, "Yao," *Zhouyi jingzhuan*, 118, lines 18a–18b.

115. Chen Jian, "Yao," *Zhouyi jingzhuan*, 116, line 16b. The manuscript uses a rare character consisting in two horizontal strokes penetrated by a single vertical stroke for *qi shi* 七十 (seventy).

116. Chen Jian, "Yao," *Zhouyi jingzhuan*, 112, line 8b. Only the lower left portion of the character transcribed as *wu* 無 (without) is visible and its identification remains uncertain. I follow Chen and Liao, "'Yao' shiwen," 434 in transcribing the final character as *zhi* 之 (it). Chen Jian notes that Zhang supplies the word *de* 德 (virtue) here.

117. Transcriptions are drawn from Xu Baogui 徐寶貴 and Wu Kejing 鄔 可晶, "Wuxing" 五行, in Qiu, *Changsha Mawangdui*, 4: 57–96. For the broader significance of the text in early Chinese thought, as well as annotated translations and transcriptions of versions found at both Mawangdui and Guodian, see Mark Csikszentmihalyi, *Material Virtue: Ethics and the Body in Early China* (Leiden: Brill, 2004).

118. Xu and Wu, "Wuxing," 58, lines 1/170–4/173; cf. Csikszentmihalyi, *Material Virtue*, 312–13, entries C 1.1–1.7.

119. See Csikszentmihalyi, *Material Virtue*, 365–66, entries C 23.1–23.3.

120. Xu and Wu, "Wuxing," 93, line 174/343; cf. Csikszentmihalyi, *Material Virtue*, 367–68, entry C 23.4. See also Pang Pu, *Zhubo "Wuxing" pian jiaozhu ji yanjiu* 竹帛〈五行〉篇校注及研究 (Taibei: Wanjuan lou, 2000), 83–84.

121. Xu and Wu, "Wuxing," 93–94, line 176/345; cf. Csikszentmihalyi, *Material Virtue*, 369, entries E 24.1–24.2.

122. Xu and Wu, "Wuxing," 94, lines 181/350; cf. Csikszentmihalyi, *Material Virtue*, 371, entries C 25.4 and E 25.4.1.

Chapter 2

1. Loewe, *Biographical Dictionary*, 242–44. As Tobias Zürn argues, the ideal ruler in the *Huainanzi* maintains a relatively obscure position, hiding or erasing the traces of his words and actions. "The sage," Zürn writes in a close-reading of one passage, "is first and foremost defined by a series of actions he does not do." See his "The Practice of Erasing Traces in the *Huainanzi*," in *The Craft of Oblivion: Forgetting and Memory in Ancient China*, ed. Albert Galvany (Albany: SUNY Press, 2023), 182.

2. Loewe, "Former Han Dynasty," 169.

3. For Liu An's legal troubles in 124, and the events leading to his trial for treason in 122, see Griet Vankeerberghen, *The* Huainanzi *and Liu An's Claim to Moral Authority* (Albany: SUNY Press, 2001), esp. 27–33.

4. Sima Qian's "Letter to Ren'an," recorded in the *Hanshu* (67.2725–36) and of possible later authorship, is the classic source for understanding his motivations for enduring castration and completing the *Shiji*. For current perspectives on the letter, see Stephen Durrant et al., *The Letter to Ren An & Sima Qian's Legacy* (Seattle: University of Washington Press, 2016).

5. The most extensive Western language study of the *Huainanzi* "Tianwen" treatise to date is Marc Kalinowski's *Maître de Huainan: Traité des Figures Célestes* (Paris: Les Belles Lettres, 2022). In his introduction, some 166 pages in length, Kalinowski marshals knowledge gained from archaeological and epigraphic work on Mawangdui, the tomb of the Marquis of Ruyin 汝陰侯 at Shuanggudui 雙古堆 (sealed in 165 BCE), and other sites to produce detailed descriptions and analyses of technical materials in the treatise, and to reconsider its organization. Nevertheless, John Major's *Heaven and Earth in Early Han Thought: Chapters Three, Four, and Five of the* Huainanzi (Albany: SUNY Press, 1993) remains a valuable study of the treatise that effectively contextualizes it within a three-chapter unit in the *Huainanzi* that also includes treatises on "Terrestrial Forms" (Dixing 墬形) and "Seasonal Rules" (Shize 時則).

6. I follow Pankenier's rendering of Shetige as Assistant Conductors (*Astrology and Cosmology*, 463). These are a pair of small constellations near Arcturus.

7. Harmonics and calendrics are the subject of two separate treatises in *Shiji* 25 and 26, while *Hanshu* 21 merges the two.

8. See Tao Lei 陶磊, "*Huainanzi Tianwen*" *yanjiu: cong shushu de jiaodu* 『淮南子·天文』研究——從數術的角度 (Jinan: Qi Lu shushe, 2003), 73–86, 119–30, 158–63.

9. Pankenier, *Astrology and Cosmology*, 302–5.

10. Boxes 2.1 and 2.2 are drawn, with some modifications, from headings included with complete annotated translations of the two treatises in Kalinowski, *Maître de Huainan*, clxxi–clxxiii and Pankenier, *Astrology and Cosmology*, 444–511. I omit the closing paragraph for the *Shiji* "Celestial Offices" as this seems to be misplaced, or possibly, an interpolation (see Pankenier, 511, n.334).

11. For a more detailed, side-by-side comparison of the technical features of the two treatises, see David Pankenier, "The *Huainanzi*'s 'Heavenly Patterns' and the *Shiji*'s 'Treatise on the Celestial Offices': What's the Difference?" in *The* Huainanzi *and Textual Production in Early China*, eds. Sarah A. Queen and Michael Puett (Leiden: Brill, 2014), 199–224, especially 207–9.

12. The inclusion of reduction (*sha* 殺) in this list suggests that the ruler must keep to cosmic cycles, though its precise sense is not entirely clear. It may mean that when plants begin to wither in the cool autumn weather, austerity is in order at court.

13. The phrase "Four Directions" occupies the same structural position as a season in the section on Saturn. This is similar to the use of "the Center" as a seasonal correspondence for Saturn in the Mawangdui "Five Planets" manuscript.

14. Two clarifications should be made concerning Saturn's correspondences: First, regarding the use of "Dipper" to determine the planet's position, the Western Jin (265–316) commentator Jin Zhuo 晉灼 (n.d.) suggests that this refers not to the Northern (i.e., Big) Dipper, but the lodge Dipper, in Sagittarius, adjacent to the asterism Establishment (Jian 建): "Regularly beginning its cycle between Establishment and Dipper in the *jiachen* year (41st in the sexagenary cycle), every year the Queller moves through a single lodge and in twenty-eight years completes its circuit of the heavens" 常以甲辰之元始建斗, 歲鎮一宿, 二十八歲而周天 (*Shiji* 27.1319, n.c1). Second, in contrast to the other planets, the punishment associated with Saturn is not included in the introduction to the planet, but several lines after it. The text associates Saturn with "the Yellow God, the favor/virtue of the ruler, and the image of the female ruler" 黃帝, 主德, 女主象也 (*Shiji* 27.1319).

15. John S. Major, Sarah Queen, Andrew Meyer, and Harold D. Roth trans., *The Huainanzi: A Guide to the Theory and Practice of Government in Early Han China* (New York: Columbia University Press, 2010), 56. Translations of passages from the *Huainanzi* are generally drawn from Major et al., a highly reliable and felicitous work, albeit with some modifications.

16. Major et al., 1.

17. Weaving is a central metaphor in the *Huainanzi* used both to describe the structure of the cosmos and the compilation of texts from multiple sources. See Tobias Benedikt Zürn, "The Han Imaginaire of Writing as Weaving: Intertextuality and the *Huainanzi*'s Self-Fashioning as an Embodiment of the Way," *Journal of Asian Studies* 79, no. 2 (2020): 367–402.

18. For a reading of the *Huainanzi* and its engagement with the views of Dong Zhongshu, an advocate of increased centralization who at times called for the execution of upstart members of collateral branches of the imperial family, see Wang Aihe, *Cosmology and Political Culture in Early China* (Cambridge: Cambridge University Press, 2000), 190–99.

19. Major et al., 113.

20. The "Summary of the Essentials" does not explicitly supply the agent who employs the *tianwen* text. However, given that the *Huainanzi* is presented to the ruler as a guide for governance, the emperor is, at least ritually and symbolically, the agent who employs the techniques and knowledge included in the treatise.

21. It is tempting to render *li* as either "set to order" or "pattern oneself on." The former rendering stresses the ruler's influence on the celestial sphere, while the latter emphasizes the imperative that the emperor treat the motions of celestial bodies as a model for his own rulership. Both valances operate simultaneously, and thus I have chosen the somewhat nebulous "pattern."

22. Liu An et al. comp., Gao You comm. *Huainanzi*, 21.370, in vol. 7 of *Zhuzi jicheng*. Citations give *juan* and page numbers; refer to citations of Major et al. for convenient section divisions. The differences between my own translation for this passage and that of Major et al. 21.2, 851 hinge on subtle distinctions in the renderings of particular words. Major et al. translate the passage in a manner that suggests the ruler acts upon the cosmos rather than accommodates himself to it. Where I render *he* 和 as "be in harmony with," for example, Major et al. render *he* as a full transitive verb, "harmonize." This is also true of our respective understandings of the verb *jie* 節, which Major et al. read as "regulate," but I understand as meaning "to follow the rhythms of" the seasons.

23. Modified from Major et al. 3.3, 117; *Huainanzi* 3.36.

24. Or, as Major et al. have it, "pure and bright" (3.1, 114).

25. Slightly modified from Major et al. 3.1, 115; *Huainanzi* 3.35.

26. Major et al. 3.2, 116; *Huainanzi* 3.35–36.

27. Slightly modified from Major et al. 3.2, 116; *Huainanzi* 3.36.

28. Cf. Major et al. 3.2, 116; *Huainanzi* 3.36.

29. Modified from Major et al. 3.16, 124; *Huainanzi* 3.40.

30. Major et al. 3.2, 115–16; *Huainanzi* 3.35.

31. Major et al. 3.22, 128; *Huainanzi* 3.43.

32. Major et al. 3.23, 129; *Huainanzi* 3.43.

33. Cf. Major et al. 3.3, 117; *Huainanzi* 3.36. Cf. Kalinowski, "Les intentions profondes du souverain s'élèvent et se communiquent au ciel" (*Maître de Huainan*, 5).

34. Cf. Major et al. 7.2, 242; *Huainanzi* 7.100.

35. Modified from Major et al. 7.2, 242–43; *Huainanzi* 7.100.

36. *Huainanzi* 20.347. Major et al. have "embraces his Heavenly Heart" (20.3, 797).

37. Major et al. 20.1, 795; *Huainanzi* 20.347.

38. Reading *feng* 風 (winds) for *feng* 鳳 (phoenixes).

39. Modified from Major et al. 20.3, 797; *Huainanzi* 20.347–48. The texts cites "Shi mai" 時邁 ("He Goes"; Mao 273). For the full ode, including early and early medieval commentary, see Kong Yingda et al. comp., *Mao shi zhengyi* 毛詩正義, 19b.320c–21c in *Shisanjing zhushu*, vol. 1. Translations of this ode and the others cited here may be found by Mao numbers in Arthur Waley trans. and Joseph R. Allen ed., *The Book of Songs: The Ancient Chinese Classic of Poetry* (New York: Grove Press, 1996). I follow this translation in rendering the titles of individual odes.

40. Major et al. 20.3, 797; *Huainanzi* 20.348. The text cites "Zheng yue" 正月 ("The First Month"; Mao 192) in support of this claim: "The *Odes* says, 'In the first month, frost is abundant; my heart is anxious and grieved'" 《詩》曰: 正月繁霜, 我心憂傷 (*Mao shi zhengyi* 12a.173c).

41. Modified from Major et al. 20.3, 797; *Huainanzi* 20.348.

42. Major et al. 12.28, 460–61; *Huainanzi* 12.199–200. Major et al. note that Zi is a surname here (460, n.100).

43. For use of this narrative in memorials, see chapter 4, pp. 131–38. It is also included in the *Shiji*, where Mars moves three degrees rather than three lunar lodges, a figure no doubt more plausible to Sima Tan and Sima Qian, both experts in mathematical astronomy.

44. Modified from Major et al. 15.8, 592; *Huainanzi* 15.257.

45. Liu Xiang comp., *Shuoyuan* 說苑, 13.15a–b, in *Siku quanshu*. This statement occurs in advice Hu Yan 狐偃 (a.k.a. Jiu Fan 咎犯) gave to Duke Wen of Jin prior to the Battle of Cheng'pu 城濮 in 632 BCE. Duke Wen likewise confronted a broom-star with its handle pointing toward his opponent, the state of Chu, but emerged victorious.

46. This is especially remarkable when compared to speeches attributed to Qi 啟, son of the Xia founder, and Tang 湯, the Shang founder, included in the *Documents* as the "Gan shi" 甘誓 (Oath at Gan) and "Tang shi" 湯誓 (Oath of Tang) chapters. Both Qi and Tang offer rewards to the valiant but threaten to execute those who avoid battle, along with their families.

47. Major et al. present a reading of the passage that differs significantly from my own. They understand this series of signs as "a sequence of bad auguries

for the endeavor of King Wu, despite which he prevails because of his superior Moral Potency" (592, n.30).

48. The *Shiji* was a private work, initiated by Sima Qian's father, that included much more than omens: (1) "Benji" 本記 (Basic Annals) that chronologically narrated major events from high antiquity through the reign of the sitting emperor; (2) "Biao" 表 (Tables); (3) "Shu" 書 (Treatises) on subjects ranging from rituals to canals; (4) "Shijia" 世家 (Hereditary Houses) chapters that told the stories of powerful families; and (5) "Lie zhuan" 列傳 (Arrayed Biographies) that related the lives of prominent people, and in many cases, preserved their writings. Though Sima Qian's official duties certainly did not entail the composition of such a history, he could not have completed the work without access to imperial archives.

49. Two interpretations of the term *wu jia* 五家 obtain: first, according to Sima Zhen's 司馬貞 (early 8th cent. CE) *Shiji suoyin* 史記索隱 (Commentary on the *Records of the Senior Archivist*), these are experts in the Five Cycles (*wu ji* 五紀) outlined in the "Great Plan" chapter of the *Documents*: the cycles of Jupiter (*sui* 歲), the moon, the sun, planets/stars and constellations (*xing chen* 星辰), and calendrical calculations (*li shu* 曆數). Zhang Shoujie's 張守節 (8th cent.) *Shiji zhengyi* 史記正義 (The Correct Meanings of the *Records of the Senior Archivist*; preface dated 737 CE) identifies the *wu jia* as five sage-kings: The Yellow God 黃帝, Gaoyang 高陽, Gaoxin 高辛, Tang Yao and Yu Shun 唐虞堯舜 (*Shiji* 27.1343, n.1–2). Rendering the phrase as Five Houses, I follow Zhang's interpretation.

50. These two lines are a quotation from the "Appended Statements" to the *Changes* (*Zhouyi zhengyi* 8.74b).

51. The Three Luminaries are the sun, moon, and stars.

52. *Shiji* 27.1342.

53. Loewe, *Biographical Dictionary*, 486.

54. Sima Qian sought to commemorate his own family line by composing the *Shiji*. See Michael Nylan, "Sima Qian: A True Historian," *Early China* 23/24 (1998–99): 245.

55. *Shiji* 27.1349. Little information concerning these three figures is otherwise available. The *Shiji* records that Tang Du was summoned to court to consult on "dividing up sections of the sky" (*fen qi tian bu* 分其天部) early in reign of Emperor Wu (26.1260). Wang Shuo commented in one instance on a comet (12.477, 28.1399), on *qi* around the sun in another (27.1338), and once admonished the general Li Guang 李廣 (d. 119) for having killed eight hundred surrendered Qiang 羌 prisoners (49.2874). Elsewhere, the "Celestial Offices" details the method Wei Xian used to read signs concerning the year. This involved identifying correspondences between wind direction at dawn on the year's first day and subsequent weather events, warfare, and disease (26.1340).

56. This event dates to 644 BCE (*Shiji* 27.1345, n.4).

57. Zhao Qi 趙岐 (d. 201 CE) identifies the Wubo as the Five Hegemons: Duke Huan 桓 of Qi (r. 685–643), Duke Wen 文 of Jin (r. 636–628), Duke Mu 穆 of Qin (r. 659–621), Duke Xiang 襄 of Song (r. 650–637), and King Zhuang 莊 of Chu (r. 613–591) (*Shiji* 27.1347, n.6).

58. This occurred in 379 BCE (*Shiji* 27.1345, n.8).

59. In 376 BCE, the rulers of Wei 魏, Han 韓, and Zhao 趙 divided up the little territory still under the control of the domain of Jin (*Shiji* 27.1346, n.9).

60. *Shiji* 27.1345. Lisa Raphals notes the consistency between the reading of cloudy *qi* in Sima Qian's treatise and the Mawangdui "Miscellaneous Sign-Readings" (*Divination and Prediction*, 349).

61. *Shiji* 27.1344.

62. *Shiji* 27.1348. Eastern Well corresponds to a group of stars in Gemini.

63. *Shiji* 27.1348.

64. *Shiji* 27.1348.

65. *Shiji* 27.1351.

66. Notwithstanding, Sima Qian points to the development of new knowledge concerning the heavens over time, recalling earlier traditions that acknowledged only the retrograde movements of Mars and not the other planets (*Shiji* 27.1349).

67. *Shiji* 1.15.

68. *Shiji* 1.15. For the closely corresponding passage in the "Canon of Yao," see Gu Jiegang 顧頡剛 (1893–1980) and Liu Qiyu 劉起釪 (1917–2012) comms., *Shangshu jiaoshi yilun* 尚書校釋譯論 (Beijing: Zhonghua, 2005) 1: 2.

69. *Shiji* 1.16–17.

70. "Yao dian," in *Shangshu jiaoshi yilun* 1: 32.

71. *Shiji* 1.24.

72. Cullen reads *qi zheng* as "the seven (concerns of) government." See Christopher Cullen and Anne S. L. Farrer, "On the Term *hsüan chi* and the Flanged Trilobate Jade Discs," *Bulletin of the School of Oriental and African Studies* 46, no. 1 (1983), 57.

73. *Shiji* 1.24.

74. Even before Sima Qian, Han thinkers saw Shun as an observer of celestial signs. Fu Sheng 伏生 (d. after 156 BCE) claimed that "Agate Orb" referred to the northern celestial pole, the point around which all stars revolve. Around the first century BCE, the armillary sphere was developed, and by the Eastern Han, it had become an essential tool for observers of celestial signs. Ma Rong interpreted *xuanji yuheng* as referring to this instrument, and Zheng Xuan claimed that the Seven Administrators referred to the sun, moon, and Five Planets (*Shiji* 1.24, n.1). For the significance of the phrase *xuanji yuheng* according to Han commentators, see Cullen and Farrer, "On the Term *hsüan chi*," 57–59.

75. *Shiji* 1.30.

210 | Notes to Chapter 2

76. The classic account of the tension between rule by virtue and rule by hereditary right is Sarah Allan, *The Heir and the Sage: Dynastic Legend in Early China* (San Francisco: Chinese Materials Center, 1981), recently republished in a revised and expanded edition (Albany: SUNY Press, 2016).

77. *Shiji* 4.117.

78. *Zhan* may refer to any kind of truncation, including beheading, amputation of a limb, or being socially "cut off" and isolated.

79. *Shiji* 4.120. Pankenier suspects that King Wu responded to a planetary omen, as Jupiter began to move retrograde in late 1048 BCE. See *Astrology and Cosmology*, 198.

80. *Shiji* 4.122, 3.108.

81. *Shiji* 4.122. This is the speech known to Sima Qian as the "Tai shi" 太誓 (Great Oath).

82. *Shiji* 2.84. Zheng Xuan writes: "The *wuxing* are the means of conducting governance so that the four seasons reach the full abundance of their power. 'To imperiously pollute them' is to violate and run counter to them. 'The Three Rectifications' are the Rectified Ways of the Celestial, Terrestrial, and Human Realms" 五行，四時盛德所行之政也。威侮，暴逆之。三正，天、地、人之正道 (*Shiji* 2.84, n.5).

83. *Shiji* 2.84.

84. The text does not say whether Xi and He refer to the same four individuals specified in the "Canon of Yao," who at this point would have been several hundred years old. Xi and He should likely be understood as lineages or clans rather than as individuals.

85. *Shiji* 2.85.

86. This campaign is presumably the subject of the "Yin Zheng" 胤征 (Campaign of Yin) text. A work under this title is included among the archaic script chapters of the received *Documents*, but it is not clear that any material included in the received chapter would have been familiar to readers prior to the 4th century CE when the archaic script chapters were compiled.

87. *Shiji* 2.85–86, 3.95.

88. Reading *li* as *lai* 賚 (reward). See *Shiji* 3.96, n.8.

89. *Shiji* 3.95. This passage corresponds in part to the received "Tang shi" chapter of the *Documents*. See *Shangshu jiaoshi yilun* 2: 883–84.

90. There is a slight incongruity between the *Shiji* passage and a quotation of Jie's speech in a fragment of the *Shangshu dazhuan* 尚書大傳 (Great Tradition of the *Venerable Documents*) cited in Pei Yin's 裴駰 (fl. 438 CE) *Shiji jijie* 史記集解 (Collected Commentaries to the *Records of the Senior Archivist*). The *Shiji* version analogizes Jie to the sun, while in the *Shangshu dazhuan* version, Jie is analogous to Heaven, and the sun is analogous to the people: "Jie said, 'Heaven has dominion over the sun, just as I have dominion over my subjects. Will the sun ever pass away? Only when the sun passes away shall I pass away'" 桀云「天

之有日, 猶吾之有民, 日有亡哉, 日亡吾亦亡矣」 (*Shiji* 3.95, n.7). In the *Shangshu dazhuan* version of Jie's speech, only when Heaven no longer holds dominion over the sun will Jie lose dominion over his people. Both versions indicate that Jie intends to rule forever.

91. The names of the Shang kings are preceded by the character *di* 帝, which I elsewhere render as "god," or, when it refers to imperial-period rulers, as "emperor." For an analysis of the possible prehistoric meanings of the word *di*, see Pankenier, "Looking to the Supernal Lord," in *Astrology and Cosmology*, 101–13.

92. An omen likewise prompted the restoration under King Taiwu. When a strange hybrid of a grain-producing plant and a mulberry tree grew to full size in a single day, Yi Zhi 伊陟 asked King Taiwu: "I have heard that monstrosities do not trump virtue (*de*). Could it be that there is any fault in your governance" 臣聞妖不勝德, 帝之政其有闕與? King Taiwu responded by cultivating his virtue, and the inauspicious tree soon withered and died (*Shiji* 3.100)

93. Zuji could equally be rendered "Ancestor Ji."

94. *Shiji* 3.103. For the corresponding section of the *Documents*, see *Shangshu jiaoshi yilun* 2: 992–1046.

95. *Shiji* 3.103. Alternatively, we might read "Way of Yin" as "Great Way," as *yin* 殷 is both a name for the Shang dynasty and an adjective meaning "great" or "large."

96. See chapter 4, pp. 122–26.

97. Cf. William H. Nienhauser et al., *The Grand Scribe's Records* (Bloomington: Indiana University Press, 1994), 1: 49.

98. *Shiji* 3.104. Alternatively, this final line might be rendered, "He commanded (*ming* 命) [his men] saying: 'Shoot Heaven!'"

99. *Shiji* 3.104.

100. *Shiji* 3.105.

101. *Shiji* 3.105. For a recent study of the figure of the *femme fatale* in the Chinese tradition and its Greek counterparts, see Yiqun Zhou, "Helen and Chinese Femmes Fatales," in *Ancient Greece and China Compared*, eds. G.E.R. Lloyd and Jingyi Jenny Zhao, in collaboration with Qiaosheng Dong (Cambridge: Cambridge University Press, 2018), 234–56.

102. See *Shangshu jiaoshi yilun* 2: 1047–70. This speech has much in common with that delivered by Zuji to Wuding earlier in the dynasty in the "Gaozong rong ri" chapter.

103. *Shiji* 3.107.

104. *Shiji* 3.107.

105. *Shiji* 4.141–42.

106. The two Co-Chancellors are not to be confused with their better-known predecessors, also called the Duke of Shao and the Duke of Zhou, who lived in the early Western Zhou.

107. *Shiji* 4.144.

108. *Shiji* 4.145.
109. *Shiji* 4.145.
110. Qi Mountain is the site to which the Zhou progenitor Gugong Danfu 古公亶父 escaped with his people to avoid war with the Xunyu 薰育, Rong 戎, and Di 狄 groups that had laid siege to their ancestral homeland, Bin 豳. See *Shiji* 4.113–14.
111. *Shiji* 4.146.
112. E.g., these sign-reading techniques required precise measurement of the movements of celestial bodies over time, a thorough understanding of the correspondences between celestial fields and locations within the terrestrial realm, and extensive knowledge of the taxonomic properties of celestial bodies.

Chapter 3

1. Mark Csikszentmihalyi, "Confucius and the *Analects* in the Han," in *Confucius and the Analects: New Essays*, ed. Bryan van Norden (Oxford: Oxford University Press, 2002), 148–49.
2. Loewe, "Former Han," 187–92.
3. *Hanshu* 36.1950.
4. Medieval historian Liu Zhiji 劉知幾 (661–721) claimed that "the 'Wuxing' was derived from Liu Xiang's work on the 'Great Plan' chapter of the *Documents*" 五行出劉向洪範. See his *Shi tong* 史通 3.8, in *Siku quanshu* 四庫全書, Wenyuange edition, intranet version (2007). The *Hanshu* bibliographic treatise does not list the title "Hongfan Wuxing zhuan lun" but credits to Liu Xiang a work entitled "Wuxing zhuan ji" 五行傳記 (Records Concerning the Traditions on the Wuxing) in eleven *juan* 卷 (fascicles), rather than eleven *pian* (*Hanshu* 36.1950, 30.1705).
5. Kongzi realized that his sagely method of governance would not be instituted in his own lifetime and declared "My way is at an end" 吾道窮矣 when the mythic unicorn (*lin* 麟) was captured by a firewood gatherer in 481 BCE. The legendary event, recounted in the final passages of the *Annals*, marks the conclusion of the Chunqiu period. See He Xiu 何休 (129–182) comm., Xu Yan 徐彥 (Tang) subcomm., *Chunqiu Gongyang zhuan zhushu* 春秋公羊傳注疏, 28.158b–160b, in vol. 2 of *Shisanjing zhushu*.
6. Michael Nylan, "On the *Hanshu* 'Wuxing zhi' 五行志 and Ban Gu's Project," in Csikszentmihalyi and Nylan, *Technical Arts*, 214. For a detailed analysis of the issues surrounding the dating and compilation of the treatise, including recent scholarly debates, see ibid. 238–44.
7. For the compilation of the "Tianwen zhi," see *Hou Han shu* 84.2784–85.
8. The interweaving of technical writing and classical authority extended beyond the *Hanshu* treatises. Two key areas for further research are the apocrypha

and later cognate treatises. The apocrypha, much like commentaries, take on titles of the Classics but concern a wide range of subjects that are not discussed, or discussed very obliquely, in the texts to which they are attached. While at least some can be dated to the Western-Eastern Han transition, the apocrypha survive only in fragments, often in much later sources, and most cannot be reliably dated. Jack L. Dull's "A Historical Introduction to the Apocryphal (Ch'an-wei) Texts of the Han Dynasty" (PhD diss., University of Washington, 1966) remains an excellent historical introduction to the topic, while recent work by Grégoire Espesset provides a methodological model for (partially) reconstructing and contextualizing these texts. See, e.g., his "Epiphanies of Sovereignty and the Rite of Jade Disc Immersion in Weft Narratives," *Early China* 37 (2014): 393–443. For later cognate treatises concerning *wuxing* and *tianwen* (or *tianxiang* 天象 [celestial images])—especially early medieval developments—two volumes are particularly recommended. B. J. Mansvelt-Beck, *The Treatises of Later Han: Their Author, Sources, Contents, and Place in Chinese Historiography* (Leiden: Brill, 1990) and Daniel P. Morgan and Damien Chaussende eds., with the collaboration of Karine Chemla, *Monographs in Tang Official Historiography: Perspectives from the Technical Treatises of the* History of the Sui (Sui shu) (Cham, Switzerland: Springer, 2019). See especially Morgan, "Heavenly Patterns," 143–79, and Nylan, "The 'Treatise on the *Wuxing*' (*Wuxing zhi*)," 181–233, in the latter volume.

9. *Hanshu* 27A.1316.

10. Dong Zhongshu's dates are typically given as ca. 179–ca. 104. For a discussion of why an earlier date of birth is more plausible, see Sarah A. Queen, *From Chronicle to Canon: The Hermeneutics of the* Spring and Autumn, According to Tung Chung-shu (Cambridge: Cambridge University Press, 1996), 241–46.

11. Su Dechang 蘇德昌 notes that the treatise includes 120 individual citations of Liu Xiang, as well as thirty-one citations in which his views are conflated with those of Dong Zhongshu. Su suspects that citations of Dong Zhongshu generally refer to a "record of disasters and anomalies" (*zai yi zhi ji* 災異之記 [possibly a title]) mentioned is his *Shiji* biography (121.3128). Eight cases indicate general agreement between Liu and Dong. In thirty entries, Liu and Dong's views are presented separately. Jing Fang 京房 the Younger's (77–37) *Jing Fang Yi zhuan* 京房易傳 (Jing Fang Tradition of the *Changes*) is another major source, cited some seventy times in the treatise. See Su Dechang, *Hanshu "Wuxing zhi" yanjiu* 漢書五行志研究 (Taipei: Taida, 2013), 130–32, 137.

12. *Hanshu* 27A.1317.

13. The phrase *huangji* could be understood in multiple ways. Nylan notes that *huang* meant "great," "sovereign," and "august," glossed as either *da* 大 or *wang* 王, while *ji* meant "to maximize," "absolute," "culminant," or "standard," as in the sense of "highest principle." The phrase as a whole meant "Greatness Maximized" or "Sovereign Standard." See Michael Nylan, *The Shifting Center:*

The Original "Great Plan" and Later Readings (Nettetal, Germany: Steyler Verlag, 1992), 17, 23–24.

14. The divisions given in this table are based on Tomiya Itaru 冨谷至 and Yoshikawa Tadao 吉川忠夫 eds., *Kanjo gogyōshi* 漢書五行志 (Tokyo: Heibonsha, 1986), ii–iv.

15. This tradition is sometimes identified as Fu Sheng's *Shangshu dazhuan* but more likely represents the work of Xiahou Shichang 夏侯始昌 (d. after 92 BCE) and later *Documents* experts to whom his teachings were transmitted. See Nylan, "On the *Hanshu* 'Wuxing zhi,'" 239.

16. *Hanshu* 27D.1458.

17. *Hanshu* 27D.1458. Ban Gu cites the phrase "for the arrogant dragon there is much humiliation" 亢龍有悔 from the line statement to the upper nine in the Qian hexagram, as well as commentary on that phrase from the "Appended Statements" (*Zhouyi zhengyi* 1.2a and 7.67c–68a).

18. This phrase is found in the "Wenyan" 文言 (Commentary on the Words) comment to the nine in the fifth place in the *Qian* hexagram (*Zhouyi zhengyi* 1.4b).

19. This event occurred in 590 BCE.

20. *Hanshu* 27D.1458–59.

21. The *Xingzhuan* is cited a total of six times in the *Hanshu*, with three citations each in the "Celestial Patterns" and the "Wuxing Treatise." No text under this title appears in the *Hanshu* bibliographic treatise.

22. Here, *chu* 除 might also be rendered "swept away." The image of sweeping is implied by the word for the comet—i.e., "broom-star." See commentary in *Chunqiu Zuozhuan zhu* 4: 1390.

23. *Hanshu* 27E.1515.

24. *Hanshu* 27E.1515–16.

25. Great Horn corresponds to Arcturus (α Boötis).

26. Translation follows Yan Shigu's suggestion that *bo* 伯 be read *ba* 霸 (*Hanshu* 27E.1516, n.1).

27. *San Qin*, or the three districts of Qin, is the collective name given to Yong 雍, Di 翟, and Sai 塞, areas roughly corresponding to modern western, northern, and eastern Shaanxi respectively, following the destruction of Qin in 206 BCE. See *Shiji* 6.275.

28. For Xiang Yu's massacre of surrendered Qin troops, said to number more than 200,000, see *Shiji* 7.310, 8.376, 92.2612; *Hanshu* 1A.44.

29. *Hanshu* 27E.1516. Liu Bang cited Yidi's murder in a speech calling for a campaign against Xiang Yu (*Hanshu* 1A.34).

30. See *Hanshu* 5.142–43.

31. Empress Chen was demoted in status in 130 BCE because she had not born a son and had been accused of committing imprecations against Wei Zifu

衛子夫 (d. 91), who was to become Emperor Wu's second empress. See Loewe, *Bibliographic Dictionary*, 31–32.

32. *Hanshu* 27E.1517.

33. For a discussion of the Banner of Chiyou in both excavated and received texts, see Michael Loewe, "The Han View of Comets" in *Divination, Mythology, and Monarchy in Han China* (New York: Cambridge University Press, 1994), 61–84.

34. This refers to campaigns initiated by Wei Qing 衛青 (d. 106) and Huo Qubing 霍去病 (ca. 140–ca. 117) against the Xiongnu. See Tomiya and Yoshikawa, *Kanjo gogyōshi*, 346, n.2.

35. See Michael Loewe, "The Fall of the House of Huo—68–66 BC" in *Crisis and Conflict in Han China 104 BC to AD 9* (London: George Allen & Unwin, 1974), 113–53, and Liu Pakyuen (Liao Boyuan) 廖伯源, Part I: "Les luttes pour le pouvoir politique au milieu de la dynastie des Han Antérieurs" (Struggles for Political Power in the mid–Western Han dynasty) of *Les institutions politiques et la lutte pour le pouvoir au milieu de la dynastie des Han Antérieurs* (Paris: Collège de France, 1983), 36–170.

36. House corresponds to stars in Pegasus.

37. For the nomination of Emperor Cheng's heir, see Griet Vankeerberghen, "Pining for the West: Chang'an in the Life of Kings and Their Families during Chengdi's Reign," in Nylan and Loewe, *China's Early Empires*, 359–60.

38. For the identification of this entry as an appearance of Halley's Comet, see Wen Shiou Tsu, "The Observations of Halley's Comet in Chinese History," *Popular Astronomy* 42 (1934): 193–94.

39. *Hanshu* 27E.1518.

40. The *Liji* does not appear to have taken its present form before the White Tiger Hall debates of 79 CE. See Jeffrey Riegel, "Li chi" in *Early Chinese Texts: A Bibliographical Guide*, ed. Michael Loewe (Berkeley: University of California, 1993), 294.

41. *Hanshu* 26.1273.

42. *Shiji* 27.1348.

43. *Hanshu* 26.1301. Cf. Ban Biao's 班彪 (3 CE–54 CE) "Wang ming lun" 王命論 (Discussion of the Charge of Kings) (*Hanshu* 100A.4211–12).

44. For omens prompting criticism, see Hans Bielenstein, "An Interpretation of the Portents in the *Ts'ien Han Shu*," *Bulletin of the Museum of Far Eastern Antiquities* 22 (1950): 127–43. For auspicious signs celebrated in hymns of praise, see Martin Kern, "Religious Anxiety and Political Interest in Western Han Omen Interpretation: The Case of the Han Wudi Period (141–87 B.C.)," *Chugoku shigaku* 中國史學 10 (Dec. 2000): 1–31.

45. This phrase is from the hexagram statement to Kun 坤 (*Zhouyi zhengyi* 1.5c).

46. *Hanshu* 26.1295–96. Basket, Axletree, and Net, respectively correspond to stars in Sagittarius, Corvus, and Taurus.

47. "Jian jian zhi shi" 漸漸之石 ("Spreading"; Mao 232). Trans. modified from Waley and Allen, *Book of Songs*, 221. For the full ode and its commentary, see *Mao shi zhengyi* 15c.231b–232c.

48. This is an elliptical quotation from the eighth division of the "Great Plan." See *Shangshu jiaoshi yilun* 3: 1187.

49. *Hanshu* 26.1295–96.

50. This appears to be a quotation but does not correspond exactly to any received text. The association between Mars and the ritual Duty of Vision is not found in any earlier astronomic treatise, nor in the "Seasonal Ordinances" chapters pertaining to summer in the *Record of Rites*.

51. *Hanshu* 26.1281. Aberrations in the movements or color of the planet manifest the penalty for lapses in the rites associated with the Duty of Vision or Summer Ordinances.

52. The section on the last month of summer in the "Monthly Ordinances" contains many sets of correspondences that had been linked to Mars since at least the early Western Han. The last month of summer is tied to Zhurong, to fire, and to the days *bing* and *ding*. However, the "Monthly Ordinances" for summer also contain a series of policy and ritual recommendations. The Son of Heaven is to reside in the right chamber of the Mingtang 明堂 (Hall of Illumination), ride in a red carriage, order fishermen to seek out submarine dragons and turtles, and command his people to labor to provide sacrifices for various deities. Should the emperor employ the wrong ordinances for the seasons, dire consequences occur. Employing the fall ordinances during the summer, for instance, produces unseasonable rains, failed crops, and disasters falling upon women in his court, or perhaps on women in the empire as a whole. See Kong Yingda et al. comp., *Liji zhengyi* 禮記正義, 16.142c–43c in vol. 2 of *Shisanjing zhushu*.

53. The phrase *tu qi* never occurs in the "Celestial Patterns" section on Saturn; however, *tu* occurs four times and is clearly associated with the planet.

54. Jiang Xiaoyuan notes that records in the "Celestial Patterns" make up approximately 30 percent of its text and take on a more standardized form than the narratives concerning early events recorded in the *Shiji* treatise. See *Tianxue zhenyuan*, 256–57.

55. Cart Ghost corresponds to stars located in Cancer.

56. *Hanshu* 26.1305.

57. *Hanshu* 26.1310.

58. In 8 BCE, four years after this occurrence, Wang Mang became Marshal of State (Loewe, *Biographical Dictionary*, 537).

59. *Hanshu* 26.1311.

60. *Hanshu* 26.1312.

61. *Hanshu* 26.1312.

62. Translation of *zhifa* 執法 after Pankenier, *Astrology and Cosmology*, 464. Pankenier suggests some uncertainty concerning the identification of the two stars (Left and Right), but he regards η and β Virginis, respectively, as most plausible.

63. *Hanshu* 26.1312.

64. Loewe, *Biographical Dictionary*, 68.

Chapter 4

1. For Mawangdui, see chapter 2, pp. 40–41, 44, 47–48.

2. *Zhouyi zhengyi* 8.7b.

3. For an overview of factional struggles in the Eastern Han, see Hans Bielenstein, "Wang Mang, the Restoration of the Han Dynasty, and Later Han," in Twitchett and Loewe, *Ch'in and Han Empires*, 274–90.

4. A succinct and useful account of the transmission of the Ouyang interpretation of the *Documents* is presented in Liang Cai, *Witchcraft and the Rise of the First Confucian Empire* (Albany: SUNY Press, 2014), 109–10.

5. *Hou Han shu* 37.1265.

6. *Hou Han shu* 37.1265.

7. *Hou Han shu* 37.1266.

8. *Hou Han shu* 37.1265.

9. Rafe de Crespigny, *A Biographical Dictionary of Later Han to the Three Kingdoms (23–220) AD* (Leiden & Boston: Brill, 2007), 415; *Hou Han shu* 82a.2717–19.

10. While Li He's call for leniency toward the Dowager Empress Yan was nominally accepted, she too died the next year (see *Hou Han shu* 10B.435–37).

11. *Hou Han shu* (treatises) 11.3242.

12. *Hou Han shu* (treatises) 11.3242, n.5. The note cites a text entitled *Li shi jia shu* 李氏家書 (Household Documents of the Li Clan). Two other references to the same text occur in the commentary to the *Hou Han shu* treatises (8.3184, n.1; 18.3364, n.2).

13. *Hou Han shu* (treatises) 11.3242, n.5.

14. *Hou Han shu* 61.2025.

15. I borrow this term from informal conversations with Robert Ashmore.

16. *Hou Han shu* 61.2025.

17. Zhou Ju specifies that it is not the ruler alone who is responsible for the drought and the underlying problems that produced it; he also blames high-status eunuchs who took wives and set up households: "It is not Your Majesty alone who has done this, but your eunuchs have also abused their power, intimidated and insulted upstanding families, and married young women, locking them away, so that they reach old age and die without ever having a partner. This is a violation of the heart of Heaven" 非但陛下行此而已，豎宦之

人, 亦復虛以形埶, 威侮良家, 取女閉之, 至有白首歿無配偶, 逆於天心 (*Hou Han shu* 61.2025).

18. *Hou Han shu* 61.2025.
19. I follow Loewe's translations for these titles. Bielenstein renders Da Sikong as "Grand Minister of Works," Da Situ as "Grand Minister over the Masses," and Da Sima as "Commander-in-chief" or "Grand Commandant." Sikou, which Bielenstein renders "Director against Brigands," was the title of an official under the Grand Marshal from 1 BCE–1 CE. For an account of the changes to the Three Excellencies in 8 BCE and thereafter, see Hans Bielenstein, *The Bureaucracy of Han Times* (Cambridge: Cambridge University Press, 1980), 11–17.
20. I read *jie* 嗟 for *cha* 差.
21. *Hou Han shu* (treatises) 13.3268, n.17. The commentary to the *Hou Han shu* ascribes the fragment of the memorial included here to Sima Biao's *Xu Hanshu* 續漢書 (Continuation of the *History of the Han*).
22. For wood and metal, see *Hanshu* 27A.1352, 1354; *Hanshu* 27B.1375; for water and fire, see *Hanshu* 27C.1405–06. For four phases disrupting soil, see *Hanshu* 27D.1441–42, 1451. In some cases, Heaven is also disrupted (see, e.g., chapter 3, p. 86). This may be why there are six disruptions rather than five.
23. *Hanshu* 27A.1353.
24. Yin Min soon lost favor due to his opposition to the apocrypha. *Hou Han shu* 79.2558.
25. Bielenstein, "Wang Mang and Later Han," 275–76.
26. De Crespigny, *Biographical Dictionary*, 1156–57.
27. *Hou Han shu* 33.1142. For the *Documents* passage, see *Shangshu jiaoshi yilun* 1: 193.
28. *Hou Han shu* 33.1141.
29. *Hou Han shu* 33.1141–42.
30. *Hou Han shu* 36.1221.
31. *Hou Han shu* 36.1221.
32. De Crespigny, *Biographical Dictionary*, 426 and 1126.
33. Zheng suggests that Yao's decision not to employ Gun 鯀, father of the sage-king Yu, resulted not from his own insights but from listening to others at his court. Zheng also cites Duke Huan of Qi's decision to employ Guan Zhong 管仲 (c. 720–645) as prime minister and Duke Wen 文 of Jin's decision to appoint Xi Hu 郤縠 (d. 632 BCE) as a military commander as instances of rulers "not being biased by their own private interests" (*bu si qi si* 不私其私). *Hou Han shu* 36.1221.
34. *Hou Han shu* 36.1222.
35. *Hou Han shu* 46.1562.
36. *Hou Han shu* 46.1563.
37. De Crespigny, *Biographical Dictionary*, 60, 840.

38. I follow Yasui Kōzan 安居香山 (1921–1989) and Nakamura Shōhachi 中村璋八 (1926–1993) in identifying these texts as "apocrypha." Fragments identified with each title may be found in their *Jōshū isho shūsei* 重修緯書集成, 6 vols. (Tokyo: Meitoku, 1971–1992). In researching this volume, I conducted a search for instances of each title in Yasui and Nakamura's survey in Han-era sources, excluding commentary. While I found more than eighty references, surprisingly few came in memorials on omens, and an analysis of the apocrypha as proof-texts must await another project, as it will require deep engagement with questions that are tangential to the present study. Of the fifty-nine references I found in the *Hou Han shu*, thirty-six occurred in the treatises, with no fewer than twenty-one in the treatise on calendrics. Many of these appeared in the writings of Jia Kui 賈逵 (31–101) and Bian Shao 邊韶 (d. 165). Only twelve occurred in the biographical chapters, including those four in Xiang Kai's memorials. The *Bohu tong* 白虎通 (Comprehensive Account of the Conference at White Tiger Hall), a text that ostensibly summarized the results of a conference in 79 CE, contained twenty-one such references. Writings attributed to Cai Yong, including memorials on omens (some duplicated in the *Hou Han shu* "Wuxing Treatise"), employed at least nine such references.

39. Rafe de Crespigny, *Portents of Protest in the Later Han Dynasty: The Memorials of Hsiang K'ai to Emperor Huan* (Canberra: Australian National University Press, 1976), 6–7.

40. *Hou Han shu* 30B.1054, 1058.

41. *Hou Han shu* 30B.1073.

42. For a brief account of this event, see Michael Loewe, "The Fall of Han," in Twitchett and Loewe, *Ch'in and Han Empires*, 317–23.

43. *Hou Han shu* 64.2117.

44. *Hou Han shu* 64.2117. I present the variant *rang* 襄 for *rang* 攘 following the Jigu Pavilion (Jigu ge 及古閣) and Wuying Hall (Wuying dian 武英殿) recensions of the text (see *Hou Han shu* 64.2128). Lu Zhi points specifically to the work of Zheng Xuan and his disciples on the "Great Plan."

45. *Shiji* 3.103. See chapter 2, p. 72, for details.

46. Alternatively, the phrase *tianyin* 天胤 may refer specifically to the ruler, or previous rulers, rather than ancestors in general (*Shangshu jiaoshi yilun* 2: 1013–14, n.3).

47. *Ni* 昵 might also refer to the ruler's close relatives rather than strictly to his father.

48. *Shangshu jiaoshi yilun* 2: 992–1011.

49. As Liu Xiang put it, King Wuding's reign was the beginning of "a hundred years of good fortune" 百年之福 (*Hanshu* 36.1964).

50. *Hanshu* 27C.1417 suggests that a group led by Wang Yin delivered the memorial. No others are named, save a certain man who, according to

Yan Shigu, bore the personal name Chong 寵, but whose surname is not given (*Hanshu* 27C.1418, n.1). Shao-yun Yang points to this memorial as the first of Emperor Cheng's reign that would be directed not at a political enemy "but to warn Chengdi against his own increasingly erratic and seemingly reckless behavior." See Yang's "The Politics of Omenology in Chengdi's Reign," in Nylan and Vankeerberghen, *Chang'an 26 BCE*, 332.

51. *Hanshu* 27C.1417.

52. *Hanshu* 27C.1418. The phrase "take control of yourself, and return to ritual propriety" is a quotation from *Lunyu* 論語 (The Analects) 12.1. See Cheng Shude 程樹德 comm., *Lunyu jishi* 論語集釋 (Beijing: Zhonghua, 1990 [preface 1939]), 817.

53. Loewe, *Biographical Dictionary*, 208.

54. The Dowager Fu is said to have died "ten or more days after" 後十餘日 the eclipse (*Hanshu* 81.3359).

55. *Hanshu* 81.3359.

56. *Hanshu* 81.3359.

57. For the *Documents* quotation, see *Shangshu jiaoshi yilun* 2: 998. The *Documents* text has *ge* 格 for *jia* 假 (came). Gu and Liu explain that *jia* is used in a number of Western Han citations of this line as the two characters had the same sound and could be used interchangeably (*Shangshu jiaoshi yilun*, 2: 1001, nn.1–2).

58. *Hanshu* 81.3359.

59. *Hanshu* 81.3360.

60. *Hanshu* 81.3360.

61. Kong Guang writes: "The exorcistic rites and petty arts of the vulgar in the end are of no benefit in responding to Heaven and blocking off the source of anomalies" 俗之祈禳小數，終無益於應天塞異 (*Hanshu* 81.3360).

62. *Hanshu* 81.3360.

63. Guan Shuxian is the immediate younger brother of King Wu (*Shiji* 35.1563).

64. I have consulted and in some instances borrowed from James Legge trans., *The Shoo King*, in *The Chinese Classics with a translation, critical and exegetical notes, prolegomena, and copious indexes*, vol. 3 in 2 parts (Hong Kong: Hong Kong University Press, 1972), 359–60 and David Nivison trans., "The Metal-Bound Coffer," in *Sources of Chinese Tradition: From Earliest Times to 1600*, eds. WM. Theodore de Bary and Irene Bloom, 2nd ed., vol. 1 (New York: Columbia University Press, 1999), 32–35.

65. *Shangshu jiaoshi yilun* 3: 1235, 1240. Punctuation has been altered to introduce a full stop after *lei dian* 雷電 (thunder and lightning).

66. *Shangshu jiaoshi yilun* 3: 1240.

67. *Shangshu jiaoshi yilun* 3: 1240.

68. *Hanshu* 85.3476.

69. *Hanshu* 85.3478. I insert a full stop after *shi zhu* 世主 (lord of the age).

70. Both Sima Qian and Wang Chong place the appearance of the violent winds and rains after the death of the Duke of Zhou (*Shiji* 33.1522; "Resonating Kinds" [Gan lei 感類] chapter of *Lun heng*). The *Shangshu dazhuan*, as cited in *Hou Han shu* 65.2141, n.4, likewise supports this view: "When the Duke of Zhou died, King Cheng wished to bury him at Chengzhou, and Heaven then sent forth thunder, rain, and wind, so that the grain stalks all lay flat upon the ground, and the great trees were uprooted. The people of the domain were very afraid" 周公薨，成王欲葬之於成周，天乃雷雨以風，禾即盡偃，大木斯拔，國人大恐.

71. Yuan Hong 袁宏 (328–376) comp., Zhou Tianyou 周天游 comm., *Hou Han ji jiaozhu* 後漢紀校注 (Tianjin: Tianjin guji, 1987), 23.642. The alternate transmission of this memorial in *Hou Han shu* 65.2141 presents a truncated version of the narrative surrounding King Cheng's posthumous treatment of the Duke of Zhou, perhaps because the story was so well known.

72. *Hou Han shu* 65.2140.

73. Huang Yi-long 黃一農 notes that, while most texts give no dates for this event, the only time when Mars moved in retrograde through Heart during Duke Jing's reign was in 494. See his *Shehui tianwenxue shi* 社會天文學史 (Shanghai: Fudan daxue, 2004), 30. The *Shiji* places the event immediately following the Chu defeat of Chen in 480, a year when Mars did not pass through Heart. The *Huainanzi*, claiming that Duke Jing died twenty-one years after the events in the narrative (451 BCE), suggests a date of 472. Plotting the location of Mars in Stellarium (v.0.22.2, stellarium.org) shows that the planet was not in retrograde motion near Heart (Scorpius) in 472, but did move in retrograde shortly before reaching Heart in March of 479, remaining so through early May, and finally passed through Heart in July and August. For an overview of Stellarium, see Georg Zotti et al., "The Simulated Sky: Stellarium for Cultural Astronomy Research," *Journal of Skyscape Archaeology* 6, no. 2 (2020): 221–58.

74. *Shiji* 38.1631.

75. Slightly modified from Major et al. 12.28, 461; *Huainanzi* 12.199.

76. Lü Buwei 呂不韋 (d. 235 BCE) comp, Gao You comm, *Lüshi chunqiu* 呂氏春秋, in vol. 6 of *Zhuzi jicheng*, j. 6, 61.

77. Jiang, *Tianxue zhenyuan*, 75–76.

78. *Hanshu* 26.1311. The *Hanshu* annals for Emperor Cheng date Zhai Fangjin's death nearly two weeks earlier, to the *renzi* 壬子 (49) day, March 14, 7 BCE. As Huang Yi-long notes, it may not be possible to resolve the conflict. See *Shehui tianwenxue shi*, 9, n.1 and *Hanshu* 10.321.

79. *Hanshu* 84.3421.

80. *Hanshu* 84.3422.

81. *Hanshu* 84.3423.

82. *Hanshu* 84.3424.

83. Huang, *Shehui tianwenxue shi*, 9.

84. *Hou Han shu* 46.1565.

85. While the people or the prime minister could perhaps be blamed as agents that had produced the sign, this was not true of the harvest. The sign itself was harmful, and Zi Wei initially implied that only the direction or object of that harm could be controlled. The conclusion to the story, in which the Dazzling Deluder moved to another celestial field, likewise suggested that its baleful influence would ultimately affect another state.

86. Li Xian's commentary specifies that the monstrous star referred to here was indeed the Dazzling Deluder (*Hou Han shu* 46.1566, n.7).

87. Emperor Cheng died the month after Zhai Fangjin (*Hanshu* 10.329).

88. Zhu Chang, at over eighty years of age, nonetheless left office near the end of 127 (De Crespigny, *Biographical Dictionary*, 1155).

89. This memorial appears in Ying Shao 應劭 (ca. 140–before 204) comp., Wang Liqi 王利器 comm., *Fengsu tongyi jiaozhu* 風俗通義校注 (Beijing: Zhonghua, 1981), 2: 5.254–55.

90. The *Hou Han shu* "Tianwen zhi" 天文志 (Treatise on Celestial Patterns) reports the movement of Mars into the Grand Tenuity constellation during the month spanning May 21, 182 through June 19, 182 (*Hou Han shu* [treatises] 20.3259). While the elongation of Venus from the sun gradually diminished from 28°37′ to 21°14′ over the course of the month, the planet would have been clearly visible in the dawn sky throughout this period. Calculations were made in Stellarium. At 151 characters in length, the extant version of the memorial may be a summary that preserves only its key points.

91. *Cai Zhonglang ji* 蔡中郎集, 2.9b–10a, in *Siku quanshu*. This memorial should be treated with some caution as Zhang Xincheng 張心澂 (1887–1973) expresses doubt regarding the authenticity of parts of Cai Yong's collected works. See *Weishu tongkao* 偽書通考 (Shanghai: Shangwu, 1939), 956.

Chapter 5

1. Modified from Michael Nylan trans., *Exemplary Figures: Fayan* (Seattle: University of Washington Press, 2013), 8.13, 129; Yang Xiong, Li Gui 李軌 comm., *Yangzi fayan* 楊子法言 (commentary compiled 335), in vol. 7 of *Zhuzi jicheng*, 8.13, 23.

2. The commentary continues, "The finest court archivists followed Heaven, and so, presented the seasons of Heaven, knew what came before and what came afterwards, and understood the state of things that obtained between the heavens and the human realm" 良史後天而奉天時, 知其所先後, 則天人之情得矣 (ibid.).

3. Schaberg, *A Patterned Past*, 96.

4. Schaberg 123–24.

5. Schaberg 97.

6. Li Wai-yee, *Readability of the Past in Early Chinese Historiography*, 195.

7. Also germane is Mark Kalinowski's work on the *Zuozhuan*. Kalinowski points to the parity between technical prediction, based on the readings of omens, and nontechnical prediction, based on the readings of human signs, such as ritual errors. Kalinowski stresses the role prediction plays in creating narrative continuity in the *Zuozhuan*, a function in part analogous to the role sign-reading sometimes plays in supporting the teleological structures of later dynastic historiography. See his "La rhétorique oraculaire."

8. While numerous omens occurred during Emperor Wu's reign that would be interpreted in the late Western Han as inauspicious, court officials during his reign generally interpreted omens as auspicious. This was especially so during the period between 113 and 103 (Kern, "Religious Anxiety," 29).

9. Reports of a given type of phenomena are not necessarily correlated to the number of actual appearances. Bielenstein notes, for example, that while reports of solar eclipses were few in number during the reign of Emperor Xuan, actual occurrences of solar eclipses were relatively frequent compared to other reigns ("An Interpretation of the Portents," 142).

10. This image of Dong largely results from the dubious attribution of the whole of the *Chunqiu fanlu* 春秋繁露 (Luxuriant Gems of the Spring and Autumn) to him. While the first seventeen *pian* (of eighty-two) may indeed have been produced by Dong or his close contemporaries, large portions of the text demonstrably belong to later periods. See Joachim Gentz, "Language of Heaven, Exegetical Skepticism and the Re-insertion of Religious Concepts in the *Gongyang* Tradition," in Lagerway and Kalinowski, *Religion in Early China*, vol. 2, 825–28; Queen, *From Chronicle to Canon*, 69–112. For a critical review of Western scholarship on Dong Zhongshu, see Michael Loewe, *Dong Zhongshu, a "Confucian" Heritage and the* Chunqiu Fanlu (Leiden: Brill, 2011), 6–18.

11. *Hanshu* 56.2495.

12. *Hanshu* 56.2496–97.

13. Michael Loewe notes that this memorial has also been dated to 140 BCE and 134 BCE (*Biographical Dictionary*, 126).

14. *Hanshu* 58.2614.

15. *Hanshu* 58.2613–14.

16. *Hanshu* 58.2614.

17. *Hanshu* 58.2614.

18. *Hanshu* 58.2614.

19. Detailed summaries and analysis of the three edicts (or rescripts) and Dong Zhongshu's responses to them are presented in Loewe, *Dong Zhongshu*, 86–100. Loewe notes that the first edict must have been composed "shortly after the accession of a young emperor, following the death of a somewhat ruthless one" (88).

20. Loewe, *Dong Zhongshu*, 121.

21. Cf. Durrant, Li, and Schaberg, *Zuo Tradition*, 7.
22. *Hanshu* 56.2501–02.
23. Dong cites "Jia le" 假樂 ("All Happiness"; Mao 249). See *Mao shi zhengyi* 17c.272c.
24. *Hanshu* 56.2505.
25. Dong cites "Da ming" 大明 ("Major Bright"; Mao 236) and "Jie Nanshan" 節南山 ("High-crested Southern Hills"; Mao 191). See *Hanshu* 56.2517, 2521.
26. Yan Shigu notes that the quotation is from the modern script "Taishi" 泰誓 (Great Oath) chapter, but it is not included in the received text, an archaic script forgery (*Hanshu* 56.2500, n.1). No fewer than three versions of the chapter existed at different points, including a lost pre-Qin version, a lost Han version, submitted to the Han court by an unnamed woman from Henan, and the archaic script version forged in the 4th century CE. See Liu Qiyu 劉起釪, *Shangshu xue shi* 尚書學史 (Beijing: Zhonghua, 1989), 33. For pre-imperial citations of the "Great Oath" not appearing in the received text, see Matsumoto Masaaki 松本雅明 (1912–1993), *Shunjū sengoku ni okeru Shōsho no tenkai: rekishi ishiki no hatten o chūshin ni* 春秋戰國における尚書の展開：歷史意識の發展を中心に (Tōkyō: Kazama Shobō, 1966), 59–62.
27. In *Shiji* 4.120, by contrast, King Wu interpreted the events as signs that it was not yet time to wage war against the Shang. See chapter 2, p. 70.
28. *Lunyu jishi* 4.25, 279. Arthur Waley translates the cited passage: "Moral force (*te*) never dwells in solitude. It always brings neighbors." See his *The Analects of Confucius* (New York: Random House, 1989 [1938]), 106.
29. *Hanshu* 56.2500.
30. *Hanshu* 56.2499.
31. *Hanshu* 56.2503.
32. According to Dong Zhongshu's biography, Gongsun Hong was jealous of his superior understanding of the *Annals*, and it was due to Gongsun Hong's suggestion that Dong was exiled to serve under the notoriously violent King of Jiaoxi 膠西 (*Hanshu* 56.2525).
33. The date of the memorial is uncertain (see note 13). The latest date suggested, 130 BCE, is the same year Gongsun Hong became Metropolitan Superintendent of the Left (Zuo Neishi 左內史). See Loewe, *Biographical Dictionary*, 126. The position of the memorial at the opening of his biography, coupled with its dismissal by the Superintendent of Ceremonial, suggests that Gongsun Hong had yet to obtain high status at court at the time of its submission.
34. Gongsun's later career, first as Metropolitan Superintendent of the Left (130), subsequently as Imperial Counsellor (126), and finally as Chancellor (124 until his death in 121), provides a sense of his ideals regarding the practical implementation of good governance. Gongsun argued unsuccessfully against maintaining military campaigns in the southwestern regions of Ba 巴 and Shu 蜀, requested that a fixed number of students attend instruction by

the Academicians, proposed a ban on the possession of bows to curb banditry, and, shortly before his death, warned of the coming rebellion of Liu An. See Loewe, *Biographical Dictionary*, 126–27.

35. Hanshu 58.2616.
36. Gongsun again repeats the language of the edict here.
37. Hanshu 58.2616.
38. Hanshu 58.2616.
39. More puzzling, however, is Gongsun's tacit attribution of severe and prolonged flooding, which Yu labored to bring under control, to the actions of Yao. Gongsun does not quite resolve the problem of baleful signs appearing under sage rulers but pushes responsibility for them from one sage to another.
40. For a detailed discussion of the contents and origins of these songs, see Martin Kern, "In Praise of Political Legitimacy: The Miao and Jiao Hymns of the Western Han," *Oriens Extremus* 38, no. 1 (1996): 29–67, esp. 48–67. The full set survives in *Hanshu* 22.1052–70.
41. I follow Yan Shigu's gloss of *xu* 緒 as *ye* 業 (*Hanshu* 22.1057, n.2). Ye refers to legacy or heritage. By extension, I interpret it to mean the occupations people inherit from their parents.
42. Zhaoyao, rendered here as "starry," is the name given to γ Boötis, the second bright star extending from the handle of the Northern Dipper. Yan Shigu held that an image of Zhaoyao was inscribed on the banner, and that such banners were carried by armies on campaign (*Hanshu* 22.1057, n.7). In its verbal or adverbial sense, *zhaoyao* describes back-and-forth motion. This line might also be rendered: "To and fro sway the banners of the Holy Ones."
43. Hanshu 22.1057.
44. Both Gu Yong and Liu Xiang objected to the construction of the Changling mausoleum. For a detailed account of the archaeology of Changling and the historical implications of the project and its abandonment, see Michael Loewe, "The Tombs Built for Han Chengdi and Migrations of the Population" in Nylan and Vankeerberghen, *Chang'an 26 BCE*, 201–17.
45. These men included three brothers and one cousin of the Dowager: Wang Feng 王鳳 33–22; Wang Yin 王音 22–15; Wang Shang 王商 15–12; and Wang Gen 王根 12–8 (Loewe, *Biographical Dictionary*, 247).
46. Palace Lady Xu (Xu Meiren 許美人) and Cao Weineng 曹偉能, the mothers of the two infants, were murdered or forced to commit suicide, apparently by Emperor Cheng's own order at the behest of his Empress Zhao Feiyan's 趙蜚燕 (d. 1 BCE) younger sister Zhao Zhaoyi 趙昭儀 (d. after 1 BCE) in 12 BCE (*Hanshu* 27B.1416, 97B.3990, 99B.3993).
47. Shao-yun Yang describes four distinct phases in the practice of reading omens over the course of Emperor Cheng's reign: "Phase 1, 32–Late 24 BCE: Omens as Weapons in Factional Conflicts"; "Phase 2, Late 24–Early 15 BCE: Omen Reports Generally Not Read in Political Terms"; "Phase 3, Early 15–

Early 12 BCE: Wangs Use Omens to Criticize Chengdi and Zhang Fang; Those Outside the Court Use Omens to Criticize Wangs"; and "Phase 4, Early 12–7 BCE: Growing Interest in Omen-Based Eschatology" ("Politics of Omenology," 323–46). The memorials under discussion here belong to the final phase, in which concern with dynastic collapse had become increasingly prevalent.

48. Liu Tseng-kuei, "Calendrical Computation Numbers and Han Dynasty Politics: A Study of Gu Yong's Three Troubles Theory," Luke Habberstad trans., in Nylan and Vankeerberghen, *Chang'an 26 BCE*, 293–322. Michael Nylan's translation of the entire memorial under discussion here, produced with the assistance of Michael Loewe, is included on pp. 296–302.

49. Shao-yun Yang points out, however, that to an extent Gu Yong treated Emperor Cheng as a victim of the times in which he lived. Nonetheless, Gu held that if Emperor Cheng were to follow his advice, crisis could ultimately be averted. Yang explains: "[A]lthough Gu Yong's interpretations still reiterated the standard omenological formula, in which good rulers receive good omens and bad rulers receive bad omens, he further claimed that the current omens carried an eschatological significance transcending any individual's virtues or misdeeds. The number of omens that had occurred since Chengdi's accession was exceptionally large, Gu Yong asserted, not because Chengdi was an exceptionally bad ruler, but because he had the misfortune to rule in exceptionally perilous times" (338).

50. *Hanshu* 85.3467.

51. The "Great Plan" points to the agricultural bounty these Five Proofs produce when they appear in balance and at the proper times, as well as the potential for harm when any one of them is absent or appears in excess. "As for the Many Proofs, these are called rain, sunshine, warmth, cold, and wind. It is said that when all five appear, each according to the proper order, the many grasses grow luxuriant. If one among them is too often present, then it is inauspicious. If one among them is too often absent, it is [also] inauspicious" 庶徵: 曰雨、曰暘、曰燠、曰寒、曰風。曰時五者來備，各以其敘，庶草蕃廡。一極備，凶; 一極無，凶 (*Shangshu jiaoshi yilun* 3: 1186–87, following Gu and Liu in reading *shi* 時 in the demonstrative pronominal sense of *shi* 是 or *ci* 此).

52. *Hanshu* 85.3467.

53. *Hanshu* 85.3467.

54. Following Nylan (Liu Tseng-kuei, "Calendrical Computation," 298).

55. *Hanshu* 85.3467. Gu's citation of the ode "Huang yi" 皇矣 ("Sovereign Might"; Mao 241) smacks of irony. Employing a variant in the latter line, the first-person pronoun *yu* 予 (translated "Our") for the prepositional *yu* 與, Gu effectively renders the ode in the collective voice of the early Zhou rulers. The Mao preface gives the gist of the ode as a whole: "'August Indeed' praises the Zhou. Heaven saw that there was none so good as Zhou to replace Yin. And among the many generations of Zhou, there was none so good as King Wen in

cultivating his own virtue" 《皇矣》、美周也。天監代殷莫若周, 周世世脩德, 莫若文王 (Mao shi zhengyi 16d.251a). Emperor Cheng is implicitly cast as falling on the wrong side of history. Another dynasty might soon come along and replace the Han. A latter-day King Wen, possessed of the virtue Emperor Cheng lacked, might soon lead an army against him. Should he fail to reform his licentious behavior and inhumane policies, ignoring all signs of danger, he might well meet the same fate that King Zhow had a millennium earlier.

56. Hanshu 85.3468.
57. Hanshu 85.3470.
58. Following Nylan (Liu Tseng-kui, "Calendrical Computation," 299).
59. Hanshu 85.3468.
60. Hanshu 85.3470.
61. Hanshu 85.3472.
62. Hanshu 85.3472.
63. Though Liu's memorial did not directly attack anyone, Ban Gu suggests that the memorial was directed against the power of the Dowager Wang and her natal family. See Hanshu 36.1966.
64. Hanshu 36.1964. Liu Xiang's treatment of celestial signs as shadows and echoes of events at court prefigures similar language in the introduction to the Hanshu, "Treatise on Celestial Patterns" (see chapter 3, p. 94).
65. Liu Xiang writes: "A careful examination of the 242 years of the Annals reveals thirty-six eclipses. These were especially numerous during the reign of Duke Xiang of Lu (r. 572–542), one eclipse occurring about every three years and five months, with a small remainder. Once the Han had flourished and become tranquil under the Filial Emperor Jing, they were [likewise] especially numerous, one eclipse occurring every three years and one month. I, your minister Xiang, have counted instances speaking of eclipses of the sun, and now, for three years running, there have been eclipses one after another. Since the Jianshi reign period (32–29), there have been eight eclipses in twenty years, occurring once every two years and six months on average" 謹案春秋二百四十二年, 日蝕三十六, 襄公尤數, 率三歲五月有奇而壹食。漢興訖竟寧, 孝景帝尤數, 率三歲一月而一食。臣向數言日當食, 今連三年比食。自建始以來, 二十歲間而八食, 率二歲六月而一發 (Hanshu 36.1963).
66. For a refutation of claims that Liu referred to a 10 BCE comet, see Christopher Cullen, "Halley's Comet and the Ghost Event of 10 BC," Quarterly Journal of the Royal Astronomical Society 32, no. 2 (1991): 113–19.
67. Purple Palace is a major constellation in the circumpolar region.
68. Hanshu 36.1965, with minor adjustments to punctuation.
69. See chapter 4, pp. 128–30.
70. This is an oft-cited quotation from the "Appended Statements" (Zhouyi zhengyi 7.70c). While it is generally used to describe the impossibility of fully conveying meaning through language, here Liu employs it to reinforce the sense

that numerous signs beyond those he explicitly mentions have occurred but not been included in his memorial. As Liu uses these lines, they might also be rendered: "One does not write down all that one would say, and one does not say all that one means."

71. Liu Xiang quotes the "Luo gao" 洛誥 (Proclamation at Luo) chapter of the *Documents*. The phrase appears in a speech made by the Duke of Zhou to King Cheng, in which he describes having sent for maps, and then performing rituals to determine if the capital should indeed be built at Luo. See *Shangshu jiaoshi yilun* 3: 1457.

72. *Hanshu* 36.1965–66.

73. For these reforms, which worked to regularize the official hierarchy, reduce conflict, and position the emperor as the unambiguous head of the Executive Council and broader government, see Luke Habberstad, "Recasting the Imperial Court in Late Western Han: Rank, Duty, and Alliances during Institutional Change," in Nylan and Vankeerberghen, *Chang'an 26 BCE*, 239–62 and Luke Habberstad, *Forming the Early Chinese Court: Rituals, Spaces, Roles* (Seattle: University of Washington Press, 2017), 124–38.

74. *Hou Han shu* 6.260.

75. *Hou Han shu* 6.259.

76. *Hou Han shu* 10B.439.

77. *Hou Han shu* 61.2021.

78. *Hou Han shu* 6.261.

79. *Hou Han ji jiaozhu* 18.507; *Hou Han shu* 6.262.

80. *Hou Han ji jiaozhu* 18.507.

81. *Hou Han shu* 59.1897–98. See also Joseph Needham (in collaboration with Wang Ling), "Astronomy" and "Seismology" in *Mathematics and the Sciences of the Heavens and the Earth*, vol. 3 of *Science and Civilization in China* (Cambridge: Cambridge University Press, 1959), 342–43, 360–61, 386, 626; and Ho, *Astronomical Chapters*, 19. For Zhang Heng's technical writings, see Yeong-Chung E Lien, "Zhang Heng, Eastern Han Polymath, His Life and Works," (PhD diss., University of Washington, 2011), UMI (3452743), 402–90, 509–33.

82. Zhang Heng surmises that the apocrypha only came into existence during the reigns of Emperor Ai and Emperor Ping 平 (r. 1 BCE–6 CE), noting an absence of earlier references to them. He attributes interest in these texts in his own day to intellectual laziness: "When it comes to musical modes and calendrics, hexagrams and observation [of *qi*?], the Nine Palaces, and wind-angle divination, these are techniques with proofs and verifications. These days, no one is willing to study them, but instead, they compete in recommending texts of no value for sign-reading. They may be compared to artisans who detest creating images of dogs and horses, but are fond of producing ghosts and spirits. Truly, reality is difficult to depict, and so, falsehood and artifice proliferate without end" 且律歷、卦候、九宮、風角，數有徵效，世莫肯學，而競稱不占之書。譬猶畫

工，惡圖犬馬而好作鬼魅，誠以實事難形，而虛偽不窮也 (*Hou Han shu* 59.1912). Zhang Heng was by no means unique in holding complex and divergent views regarding different sign-reading techniques. As Richard J. Smith has recently written, "criticisms of mantic practices . . . have almost invariably been directed toward one or another technique (or type of practitioner), not against the idea of divination itself." See Smith's "Prognostication in Premodern China: Issues of Culture and Class," in *Handbook of Prognostication in China*, Part 1, eds. Michael Lackner and Zhao Lu (Leiden: Brill, 2022), 128.

83. *Hou Han shu* 59.1940.

84. The bronze instrument consisted in eight dragon heads facing eight directions, each of which held in its mouth an iron ball. Should a quake occur in a given direction, the iron ball would issue from the corresponding dragon's mouth and fall into the mouth of a toad. In one instance, the instrument identified an earthquake in Longxi 隴西, news of which arrived by messenger several days later (*Hou Han shu* 59.1909). For a reconstruction of the instrument, see André Wegener Sleeswyk and Nathan Sivin, "Dragons and Toads: The Chinese Seismoscope of A.D. 132," *Chinese Science* 6 (Nov. 1983), 1–19.

85. *Hou Han shu* 59.1909.

86. *Hou Han shu* 6.260, 59.1909.

87. *Hou Han ji jiaozhu* 18.512.

88. *Hou Han ji jiaozhu* 18.512.

89. *Hou Han ji jiaozhu* 18.514.

90. *Hou Han shu* 6.261.

91. *Hou Han ji jiaozhu* 18.513.

92. *Hou Han ji jiaozhu* 18.513.

93. *Hou Han ji jiaozhu* 18.512.

94. *Hou Han ji jiaozhu* 18.514.

95. Ma Rong was an expert in the study of celestial patterns and, it will be recalled from chapter 3, helped to compile the *Hanshu* "Treatise on Celestial Patterns." He omitted a host of planetary portents that might have been added to the recent earthquake. Ma drew the attention of his court readership only to signs that directly affected agricultural production.

96. *Hou Han ji jiaozhu* 18.510.

97. *Hou Han ji jiaozhu* 18.510–11.

98. More literally, "the same in their culpability."

99. *Hou Han ji jiaozhu* 18.511.

100. *Hou Han ji jiaozhu* 18.511.

101. *Hou Han ji jiaozhu* 18.511.

102. For the "Eight Concerns of Governance," see chapter 4, p. 112, and box 4.1. The "Tianguan zhong zai" 天官冢宰 (High Ministers of the Celestial Offices) chapter of the *Rites of Zhou* (Zhouli 周禮) describes the management and roles of the Nine Occupations as follows: "By means of the Nine Occupations

employ the myriad people. The first includes the three kinds of farmers who raise the Nine Grains; the second are the gardeners, who cultivate the grasses and trees; the third are the foresters, who produce resources from the mountains and marshes; the fourth are the fowl-keepers and herdsmen, who nurture the birds and beasts; the fifth are the Hundred Artisans, who embellish and transform the Eight Materials; the sixth are the merchants, who circulate goods and make them plentiful; the seventh are the noble women, who work the silk and hemp; the eighth are servant men and women, who gather and harvest the various materials; and the ninth are the idle people, who are without regular employment, but go from doing one thing to doing another" 以九職任萬民：一曰三農，生九穀；二曰園圃，毓草木；三曰虞衡，作山澤之材；四曰藪牧，養蕃鳥獸；五曰百工，飭化八材；六曰商賈，阜通貨賄；七曰嬪婦，化治絲枲；八曰臣妾，聚斂疏材；九曰閑民，無常職，轉移執事 (Jia Gongyan 賈公彥 [7th cent.] et al. comp., *Zhouli zhushu* 周禮注疏 2.9a, in vol. 1 of *Shisanjing zhushu*).

103. *Hou Han ji jiaozhu* 18.511.
104. *Hou Han ji jiaozhu* 18.511.
105. *Hou Han ji jiaozhu* 18.509.
106. *Hou Han ji jiaozhu* 18.508.
107. *Hou Han ji jiaozhu* 18.508.
108. *Hou Han ji jiaozhu* 18.509.
109. *Hou Han ji jiaozhu* 18.509.
110. *Hou Han ji jiaozhu* 18.509–10.
111. *Yangzi fayan* 13.15, 41. Slightly modified from Nylan, *Exemplary Figures*, 230–31.

Conclusion

1. Latour, *We Have Never Been Modern*, 5.
2. As early as the 1930s, Robert Merton noted that the "radical cleavage between science and religion" was a relatively recent phenomenon. See his "The Puritan Spur to Science" [1938] in *The Sociology of Science: Theoretical and Empirical Investigations* (Chicago: University of Chicago Press, 1973), 233.
3. While a positivist approach to the technical works of Liu Xiang, Ma Rong, or Zhang Heng might posit that their scientific activities were separate from their politics, their classical learning, or their poetry, Paul Feyerabend cautions against any approach to the history of science in which "[a] person's religion . . . or his metaphysics, or his sense of humour . . . must not have the slightest connection with his scientific activity" (*Against Method*, 3rd ed. [London: Verso, 1993], 11).
4. As Loewe puts it, "Dong Zhongshu, a Chinese Pericles, may not himself have been as impressive a writer as his biographer, a Thucydides" (*Dong Zhongshu*, 121).

5. De Crespigny, *Biographical Dictionary*, 649. Much of this work was likely completed after his retirement from public service in 147.

6. See Liu Xiang's "Vanishing in the Distance" (Yuan shi 遠逝), *Chuci buzhu* 16.292–95, translated as "Going Far Away" in David Hawkes trans., *Ch'u Tz'u: The Songs of the South* (Boston: Beacon, 1962), 158–59; Ma Rong's "Hymn for Guangcheng" (Guangcheng song 廣成頌), *Hou Han shu* 60A.1954–69; and Zhang Heng's "Contemplating the Mystery" (Si xuan fu 思玄賦), *Hou Han shu* 59.1914–38, translated by David R. Knechtges as "Rhapsody on Contemplating the Mystery," in *Wen Xuan*, or *Selections of Refined Literature* (Princeton, NJ: Princeton University Press: 1982), 3: 105–38. For an essay on Zhang Heng's classicist adaptation of themes of the genre, see David R. Knechtges, "A Journey to Morality," in *Essays in Commemoration of the Golden Jubilee of the Fung Ping Shan Library, 1932–1982*, eds. Chan Ping-leung et al. (Hong Kong: Fung Ping Shan Library of the University of Hong Kong, 1982), 162–82.

7. Oedipus's doom in Sophocles's (ca. 497–ca. 406) *Oedipus the King*, for example, proved inescapable. Taking action to prevent the baleful events the oracle had predicted only played into the hands of fate. Reluctant to tell Oedipus of his ineluctable demise, the prophet Tiresias declaimed: "Alas, alas, how dreadful it is to know when the knowledge does not benefit the knower!" See Hugh Lloyd-Jones trans., *Sophocles: Ajax, Electra, Oedipus Tyrannus* (Cambridge, MA: Harvard University Press, 1994), 355, lines 316–18.

8. For an extended discussion of this shift in Gongyang exegesis, see Gentz, "Language of Heaven," 836–38.

Appendix A

1. In her analysis of the mass influx into Chinese of new terms drawn from European languages that occurred in the early twentieth century, Lydia Liu has shown that hypothetical equivalents for new concepts often had to be discovered or invented. Terms became meaningful in Chinese in a manner largely independent of their precise sense in their original linguistic contexts. See Liu's *Translingual Practice: Literature, National Culture, and Translated Modernity—China, 1900–1937* (Stanford, CA: Stanford University Press, 1995), 3–40.

2. For "counterparts," see Schafer, *Pacing the Void*, 5–6 and 55–56.

3. For a monograph devoted in large part to the development of *rui* in the early imperial and medieval periods, see Lippiello's *Auspicious Omens*.

4. See the discussion of the memorials of Dong Zhongshu and Gongsun Hong in chapter 5, pp. 143–53.

5. See Mark Csikszentmihalyi, "Protective Talismans" in *Readings in Han Chinese Thought* (Indianapolis: Hackett, 2006), 156–66.

6. The phrase *zaiyi* became much more common in the late Western Han. In the whole of the *Shiji*, the phrase occurs only three times (all in a

single paragraph concerning Dong Zhongshu on *Shiji* 121.3128). In the *Hanshu*, by contrast, the phrase occurs eighty-seven times.

7. Mark Csikszentmihalyi, "The Social Roles of the *Annals* Classic in Late Western Han," in *Chang'an 26 BCE: An Augustan Age in China*, eds. Michael Nylan and Griet Vankeerberghen (Seattle: University of Washington Press, 2014), 466.

8. For "alternation" as the sense of *bian*, see Nathan Sivin, "Change and Continuity in Early Cosmology: *The Great Commentary to The Book of Changes*" in Yamada Keiji 山田慶兒 ed., *Chūgoku kodai kagakushi ron zokuhen* 中国古代科学史論続編 (Kyoto: Jinbun Kagaku Kenkyūjo, 1989), 3–43.

9. See chapter 3, p. 89.

10. *Hou Han shu* 59.1909.

11. *Hanshu* 85.3467.

12. *Shangshu jiaoshi yilun* 3: 1186–87.

13. For a historical overview of the relationship between *wuxing*, *yinyang*, and *qi* in early China, see Michael Nylan, "Yin-yang, Five Phases, and *qi*," in Nylan and Loewe, *China's Early Empires*, 398–413.

14. Perhaps the most central thesis of the "Great Plan" is that "the ruler must give away in order to retain and increase" (Nylan, *Shifting Center*, 25). With regard to *wuxing*, Nylan explains: "According to the earliest commentary, the *wu-hsing* appear here simply as resources of the empire, but as their exact connection with government policy is unclear, their inclusion is somewhat puzzling. Perhaps they are to remind the ruler of his obligation to regulate certain material elements in order to provide the minimum economic base necessary for successful rule" (*Shifting Center*, 15).

15. For the identification of the tradition, see p. 216, n. 15.

16. *Hanshu* 27C.1405.

17. *The Way and Its Power*, 20.

18. Among instances in other early texts, this phrase appears in the *Documents* (*Shangshu jiaoshi yilun*, 4: 1638), twice in the *Hanshu* (52.2406, 100B.4253), and four times in a single passage in the *Zuozhuan* (*Chunqiu Zuozhuan zhu*, 2: 635–38).

Appendix B

1. *Hanshu* 30.1763–65.

2. Yao Zhenzong, *Hanshu yiwenzhi tiaoli*, 374–85.

3. The Five Remnants (*wu can*) are listed in the *Shiji* "Celestial Offices" (27.1333) along with other transient *xing* that carry auspicious or inauspicious meanings. Because these do not have fixed positions, they are best understood not as fixed stars but as meteors, comets, novae, or optical phenomena surrounding

fixed stars or planets. Pankenier renders *wu can* as the Five Cruelties Star (*Astrology and Cosmology*, 494). Ma Guohan 馬國漢 (1794–1857) compiled a collection of fragments from this text based on Meng Kang's 孟康 (3rd cent. CE) *Shiji* commentary (Yao, *Hanshu yiwenzhi tiaoli*, 374).

4. Yan Shigu claims that Chang Cong was a teacher of *Laozi* (*Hanshu* 30.1765, n.2).

5. Dongfang Shuo 東方朔 (154–93) received a promotion to the rank of Grand Counsellor of the Palace (Taizhong dafu 太中大夫) after submitting a text with this title in 138 BCE. Meng Kang notes that the *Grand Stairway* refers to the *Three Platforms* (San tai 三台), a constellation formed by three asterisms, each of which includes two proximate stars to form a single platform, all of which correspond to parts of Ursa Major. Yan Shigu's *Hanshu* commentary contains a long fragment that may have been part of this text (*Hanshu* 65.2851, n.4).

6. Jade Crossbar corresponds to ε Ursae Majoris. "Han" in the title probably refers not to the name of a dynasty but to the "Yun Han" 雲漢, the "Cloudy River" known in modern English as the "Milky Way."

7. While it is tempting to read "Wuxing jing" as "The Classic of the Five Planets," I have not found other references to a text of this title. Translation is tentative.

Works Cited

Received Sources in Classical Chinese

Texts in this category are cited by the titles of the editions in which they appear. These often include modern commentary.

Cai Zhonglang ji 蔡中郎集 (Collected Works of Cai Yong 蔡邕 [132–192 CE]). Cai Yong. In *Siku quanshu*.

Chuci buzhu 楚辭補注 (A Supplementary Commentary to the *Lyrics of Chu*). Hong Xingzu 洪興祖 (1090–1155) and Wang Yi 王逸 (fl. 89–158) comms. Beijing: Zhonghua, 1983.

Chunqiu Gongyang zhuan zhushu 春秋公羊傳注疏 (Commentary and Subcommentary to the *Gongyang Tradition* of the *Annals*). He Xiu 何休 (129–182) comm., Xu Yan 徐彥 (Tang) subcomm. In vol. 2 of *Shisanjing zhushu*.

Chunqiu Zuozhuan zhu 春秋左傳注 (Commentary to the *Zuo Tradition* of the *Annals*). Yang Bojun 楊伯峻 (1909–1992) comm. 4 vols. Beijing: Zhonghua, 1981.

Fengsu tongyi jiaozhu 風俗通義校注 (Text-critical Commentary to a *Comprehensive Account of Customs*). Ying Shao 應劭 (ca. 140–before 204) comp., Wang Liqi 王利器 comm. 2 vols. Beijing: Zhonghua, 1981.

Hanshu 漢書 (History of the Han). Ban Gu 班固 (32–92) comp., Yan Shigu 顏師古 (581–645) comm. Beijing: Zhonghua, 1962.

Hou Han ji jiaozhu 後漢紀校注 (Text-critical Commentary to *Records of the Later Han*). Yuan Hong 袁宏 (328–376) comp., Zhou Tianyou 周天游 comm. Tianjin: Tianjin guji, 1987.

Hou Han shu 後漢書 (History of the Later Han Dynasty). Fan Ye 范曄 (398–445) comp., Li Xian 李賢 (651–684) comm. Beijing: Zhonghua, 1965. Technical treatises (*zhi* 志) compiled by Sima Biao 司馬彪 (240–306) with commentary by Liu Shao 劉昭 (fl. ca. 502–ca. 519).

Huainanzi 淮南子. Liu An 劉安 (179–122) et al. comp. Gao You 高誘 (ca. 168–212) comm. In vol. 7 of *Zhuzi jicheng*.

Liji zhengyi 禮記正義 (Correct Meaning of the *Record of Rites*). Kong Yingda 孔穎達 (574–648) et al. comp. In vol. 2 of *Shisanjing zhushu*.

Lun heng 論衡 (Balanced Discourses). Wang Chong 王充 (27–ca. 100). In vol. 7 of *Zhuzi jicheng*.

Lunyu jishi 論語集釋 (Collected Commentaries to the *Analects*). Cheng Shude 程樹德 comm. 4 vols. Preface dated 1939. Beijing: Zhonghua, 1990.

Lüshi chunqiu 呂氏春秋 (Annals of Lü Buwei). Lü Buwei 呂不韋 (d. 235 BCE) comp. Gao You comm. In vol. 6 of *Zhuzi jicheng*.

Mao shi zhengyi 毛詩正義 (Correct Meanings of the Mao Odes). Kong Yingda et al. comp. In vol. 1 of *Shisanjing zhushu*.

Ruan Yuan 阮元 (1764–1849) coll. 2 vols. Beijing: Zhonghua, 1980.

Shangshu jiaoshi yilun 尚書校釋譯論 (Text-critical Notes, Translation, and Discussion of the *Venerable Documents*). Gu Jiegang 顧頡剛 (1893–1980) and Liu Qiyu 劉起釪 (1917–2012) comms. 4 vols. Beijing: Zhonghua, 2005.

Shi tong 史通 (Comprehensive History). Liu Zhiji 劉知幾 (661–721). In *Siku quanshu*.

Shiji 史記 (Records of the Senior Archivist). Sima Qian 司馬遷 (ca. 145–ca. 86) comp. Beijing: Zhonghua, 1959.

Shisanjing zhushu 十三經注疏 (Commentary and Subcommentary to the *Thirteen Classics*). Ruan Yuan 阮元 (1764–1849) coll. 2 vols. Beijing: Zhonghua, 1980.

Shuoyuan 說苑 (Garden of Anecdotes). Liu Xiang 劉向 (79–8 BCE) comp. In *Siku quanshu*.

Siku quanshu 四庫全書 (Complete Books of the Four Chambers). Wenyuange edition. Intranet version, 2007.

Xunzi jijie 荀子集解 (Collected Exegesis on the *Xunzi*). Wang Xianqian 王先謙 (1842–1918) comm. Vol. 2 of *Zhuzi jicheng*.

Yangzi fayan 楊子法言 (The Exemplary Figures of Master Yang). Yang Xiong 揚雄 (53 BCE–18 CE), Li Gui 李軌 comm. (commentary compiled 335). In vol. 7 of *Zhuzi jicheng*.

Yanzi chunqiu 晏子春秋 (Annals of Master Yan). In vol. 4 of *Zhuzi jicheng*.

Yanzi chunqiu jishi 晏子春秋集釋 (Collected Commentaries to the *Annals of Master Yan*). Wu Zeyu 吳則虞 (1913–1977) comm. 2 vols. Beijing: Zhonghua, 1962.

Zhouli zhushu 周禮注疏 (Commentary and Subcommentary to the *Rites of Zhou*). Jia Gongyan 賈公彥 (7th cent.) et al. comp. In vol. 1 of *Shisanjing zhushu*.

Zhouyi zhengyi 周易正義 (Correct Meanings of the *Changes of Zhou*). Kong Yingda et al. comp. In vol. 1 of *Shisanjing zhushu*.

Zhuzi jicheng 諸子集成 (Collected Works of the Many Masters). 8 vols. Beijing: Zhonghua, 1952.

Excavated Sources

Chen Jian 陳劍. *Zhouyi jingzhuan* 周易經傳 (*Changes of Zhou*: Classic and Traditions). In Qiu Xigui, *Mawangdui jicheng*, 3: 3–162.

Chen Songchang 陳松長. *Mawangdui boshu "Xingde" yanjiu lungao* 馬王堆帛書《刑德》研究論稿 (Summary of Research on the Mawangdui Silk "Harshness and Favor" Texts). Taibei: Taiwan guji, 2001.

Chen Songchang 陳松長 and Liao Mingchun 廖名春. "'Yao' shi wen" 《要》釋文 (Annotated Transcription of the "Essentials"). *Daojia wenhua* 道家文化 (Daoist Culture) 3 (1993): 434–35.

Cheng Shaoxuan 程少軒. "Xingde jia" 刑德甲 (Harshness and Favor A). In Qiu Xigui, *Mawangdui jicheng*, 5: 1–30.

Hirose Kunio 廣瀨薰雄. *Jing fa* 經法 (Constant Models). In Qiu Xigui, *Mawangdui jicheng*, 4: 125–50.

Liu Jianmin 劉建民. "Wuxing zhan" 五星占 (Five Planets Sign-Readings). In Qiu Xigui, *Mawangdui jicheng*, 4: 223–44.

Liu Lexian 劉樂賢. *Mawangdui tianwen shu kaoshi* 馬王堆天文書考釋 (An Examination and Interpretation of the Mawangdui Texts on Celestial Patterns). Guangzhou: Zhongshan daxue, 2004.

"Mawangdui Hanmu boshu 'Wuxing zhan' shiwen" 馬王堆漢墓帛書〈五星占〉釋文 (Exegesis on the Text of the "Five Planets Sign-Readings" Silk Manuscript from the Han Tombs at Mawangdui). In *Zhongguo tianwenxue shi wenji* 中國天文學史文集 (Collected Documents on the History of Astronomy in China), 1–13. Beijing: Kexue, 1979.

Pang Pu 龐樸. *Zhubo "Wuxing" pian jiaozhu ji yanjiu* 竹帛〈五行〉篇校注及研究 (Text-critical Commentary and Research on the Bamboo and Silk "Five Actions" Text). Taibei: Wanjuan lou, 2000.

Qiu Xigui 裘錫圭 ed. *Changsha Mawangdui Hanmu jianbo jicheng* 長沙馬王堆漢墓簡帛集成 (Collected Bamboo and Silk Texts from Han Tombs at Mawangdui, Changsha). 7 vols. Shanghai: Fudan, 2014.

Wang Shujin 王樹金. "Tianwen qixiang zazhan" 天文氣象雜占 (Miscellaneous Sign-Readings on Astronomical and Meteorological Phenomena). In Qiu Xigui, *Changsha Mawangdui*, 4: 245–75.

Xu Baogui 徐寶貴 and Wu Kejing 鄔可晶. "Wuxing" 五行 (Five Actions). In Qiu Xigui, *Mawangdui jicheng*, 4: 57–96.

Yamada Keiji 山田慶兒. "Tenmon kishō zatsusen" 天文氣象雜占 (Miscellaneous Sign-Readings on Astronomical and Meteorological Phenomena). In *Shinhatsugen Chūgoku kagakushi shiryō no kenkyū* 新發現中國科學史資料の研究 (Research on Newly Discovered Materials in the History of Science in China), 45–86. Kyōto: Kyōto daigaku jinbun kagaku kenkyūjo, 1985.

Zhang Zhenglang 張政烺. *Mawangdui boshu Zhouyi jingzhuan jiaodu* 馬王堆帛書周易經傳校讀 (A Text-Critical Reading of the Mawangdui Silk *Zhou Changes* Classic and Its Commentaries). Beijing: Zhonghua, 2008.

Translations and Secondary Sources

Allan, Sarah. "Erlitou and the Formation of Chinese Civilization: Toward a New Paradigm." *Journal of Asian Studies* 66, no. 2 (May 2007): 461–96.

———. *The Heir and the Sage: Dynastic Legend in Early China*. San Francisco: Chinese Materials Center, 1981. Recently republished in a revised and expanded edition. Albany: SUNY Press, 2016.

Astrophysikalisches Institut und Universitäts-Sternwarte. "Terra-Astronomy." Friedrich-Schiller-Universität Jena. www.astro.uni-jena.de/index.php/terra-astronomy.html.

Bielenstein, Hans. *The Bureaucracy of Han Times*. Cambridge: Cambridge University Press, 1980.

———. "An Interpretation of the Portents in the *Ts'ien Han Shu*." *Bulletin of the Museum of Far Eastern Antiquities* 22 (1950): 127–43.

———. "Wang Mang, the Restoration of the Han Dynasty, and Later Han." In Twitchett and Loewe, *Ch'in and Han Empires*, 274–290.

Bodde, Derk. "The State and Empire of Ch'in." In Twitchett and Loewe, *Ch'in and Han Empires*, 20–102.

Brown, David. *Mesopotamian Planetary Astronomy-Astrology*. Groningen: Brill, 2000.

Brown, Miranda. "Liu Xiang: The Imperial Library and the Creation of the Exemplary Healer List." In *The Art of Medicine in Early China: The Ancient and Medieval Origins of a Modern Archive*, 89–109. Cambridge: Cambridge University Press, 2015.

Cai, Liang. *Witchcraft and the Rise of the First Confucian Empire*. Albany: SUNY Press, 2014.

Chapman, Jesse J. "Ambiguity, Scope, and Significance: Difficulties in Interpreting Celestial Phenomena in Chinese Records." *Proceedings of the International Astronomical Union* 14, no. A30 (2018): 152–55.

Chapman, J., M. Csikszentmihalyi, and R. Neuhäuser. "The Chinese Comet Observation in AD 773 January." *Astronomische Nachrichten* 335, no. 9 (Nov. 2014): 964–67.

Chen Meidong 陳美東. *Zhongguo gudai tianwenxue sixiang* 中國古代天文學思想 (Ancient Chinese Astronomical Thought). Beijing: Zhongguo kexue jishu, 2008.

Chen Songchang. "Mawangdui sanhao muzhu de zai renshi" 馬王堆三號墓主的再認識 (Re-identifying the Tomb Occupant of Mawangdui Tomb 3). In

Jianbo yanjiu wengao 簡帛研究文稿 (Articles on Silk and Bamboo Research), 435–42. Beijing: Xianzhuang shuju, 2008.

Csikszentmihalyi, Mark. "Confucius and the *Analects* in the Han." In *Confucius and the Analects: New Essays*, edited by Bryan van Norden, 134–62. Oxford: Oxford University Press, 2002.

———. *Material Virtue: Ethics and the Body in Early China*. Leiden & Boston: Brill, 2004.

———. "Protective Talismans," 156–66. In *Readings in Han Chinese Thought*. Indianapolis: Hackett, 2006.

———. "The Social Roles of the *Annals* Classic in Late Western Han." In Nylan and Vankeerberghen, *Chang'an 26 BCE*, 461–76.

Csikszentmihalyi, Mark, and Michael Nylan eds. *Technical Arts in the Han Histories: Tables and Treatises in the* Shiji *and* Hanshu. Albany: SUNY Press, 2021.

Cullen, Christopher. "Halley's Comet and the Ghost Event of 10 BC." *Quarterly Journal of the Royal Astronomical Society* 32, no. 2 (1991): 113–19.

———. *Heavenly Numbers: Astronomy and Authority in Early Imperial China*. Oxford: Oxford University Press, 2017.

———. "Numbers, Numeracy and the Cosmos." In Nylan and Loewe, *China's Early Empires*, 323–38.

———. "Understanding the Planets in Ancient China: Prediction and Divination in the *Wu xing zhan*." *Early Science and Medicine* 16 (2001): 218–51.

———. "*Wu Xing Zhan* 五星占 'Prognostics of the Five Planets.'" *SCIAMVS: Sources and Commentaries in the Exact Sciences* 12 (2011): 193–249.

Cullen, Christopher, and Anne S. L. Farrer. "On the Term *hsüan chi* and the Flanged Trilobate Jade Discs." *Bulletin of the School of Oriental and African Studies* 46, no. 1 (1983): 52–76.

de Crespigny, Rafe. *A Biographical Dictionary of Later Han to the Three Kingdoms (23–220) AD*. Leiden & Boston: Brill, 2007.

———. *Portents of Protest in the Later Han Dynasty: The Memorials of Hsiang K'ai to Emperor Huan*. Canberra: Australian National University Press, 1976.

Dull, Jack L. "A Historical Introduction to the Apocryphal (Ch'an-wei) Texts of the Han Dynasty." PhD diss., University of Washington, Seattle, 1966.

Durrant, Stephen et al. *The Letter to Ren An & Sima Qian's Legacy*. Seattle: University of Washington Press, 2016.

Durrant, Stephen, Wai-yee Li, and David Schaberg, trans. *Zuo Tradition (Zuozhuan): Commentary on the "Spring and Autumn Annals."* Seattle: University of Washington Press, 2016.

Eno, Robert. "Deities and Ancestors in Early Oracle Inscriptions." In *Religions of China in Practice*, edited by Donald S. Lopez Jr., 41–51. Princeton, NJ: Princeton University Press, 1996.

Espesset, Grégoire. "Epiphanies of Sovereignty and the Rite of Jade Disc Immersion in Weft Narratives." *Early China* 37 (2014): 393–443.
Feng Shi 馮時. *Zhongguo tianwen kaoguxue* 中國天文考古學 (Archaeoastronomy in China). Beijing: Shehui kexue wenxian, 2001.
Forke, Alfred, trans. *Lun-Hêng*. New York: Paragon Book Gallery, 1962.
Fu Juyou 傅舉有. "Han dai liehou de jiali—jian tan Mawangdui sanhao mu muzhu" 漢代列侯的家吏——兼談馬王堆三號墓墓主 (The Family Servant of a Han Dynasty Marquis—A Thorough Discussion of the Tomb Occupant of Mawangdui Tomb 3). *Wenwu* 文物 (Cultural Relics) 1 (1999): 86–89, 96.
Fukui Shigemasa 福井重雅. *Kandai Jukyō no shiteki kenkyū: Jukyō no kangakuka o meguru teisetsu no saikentō* 漢代儒教の史的研究: 儒教の官學化をめぐる定說の再檢討 (Historical Research on Classicism in the Han Dynasty: A Re-evaluation of the Accepted Ideals Regarding the Officialization of Classicism). Kyūko sōsho 60. Tokyo: Kyūko Shoin, 2005.
Galvany, Albert. "Signs, Clues and Traces: Anticipation in Ancient Chinese Political and Military Texts." *Early China* 38 (2015): 1–43.
Gentz, Joachim. "Language of Heaven, Exegetical Skepticism and the Reinsertion of Religious Concepts in the *Gongyang* Tradition." In Lagerwey and Kalinowski, *Early Chinese Religion*, 814–38.
Goodman, Howard L. "Chinese Polymaths, 100–300 AD: The Tung-Kuan, Taoist Dissent, and Technical Skills." *Asia Major* 3rd ser., 18.1 (2005): 101–74.
Gu Tiefu 顧鐵符 (1908–1990). "Mawangdui boshu 'Tianwen qixiang zazhan' neirong jianshu" 馬王堆帛書天文氣象雜占內容簡述 (A Brief Description of the Content of the Mawangdui Silk Text "Miscellaneous Sign-Readings on Astronomical and Meteorological Phenomena"). *Wenwu* 2 (1978): 1–4.
Habberstad, Luke. *Forming the Early Chinese Court: Rituals, Spaces, Roles*. Seattle: University of Washington Press, 2017.
———. "Recasting the Imperial Court in Late Western Han: Rank, Duty, and Alliances during Institutional Change." In Nylan and Vankeerberghen, *Chang'an 26 BCE*, 239–62.
Harkness, Ethan. "Cosmology and the Quotidian: Day Books in Early China." PhD diss., University of Chicago, 2011. UMI (3487616).
Harper, Donald. "Communication by Design: Two Silk Manuscripts of Diagrams (*Tu*) from Mawangdui Tomb Three." In *Graphics and Text in the Production of Technical Knowledge in China: The Warp and the Weft*, edited by Francesca Bray, Vera Dorofeeva-Lichtmann, and Georges Métailié, 169–89. Leiden: Brill, 2007.
———. *Early Chinese Medical Literature: The Mawangdui Medical Manuscripts*. London: Kegan Paul, 1998.
Harper, Donald, and Marc Kalinowski, eds. *Books of Fate and Popular Culture in Early China: The Daybook Manuscripts of the Warring States, Qin, and Han*. Handbuch der Orientalistik IV.33. Leiden & Boston: Brill, 2017.

Hawkes, David, trans. *Ch'u Tz'u: The Songs of the South*. Boston: Beacon, 1962.
He Jiejun 何介鈞. *Changsha Mawangdui er san hao Hanmu* 長沙馬王堆二三號漢墓 (Han Dynasty Tombs 2 and 3 at Mawangdui, Changsha). Beijing: Wenwu, 2004.
Henderson, John B. "Divination and Confucian Exegesis." *Extrême-Orient, Extrême-Occident* 21 (1999): 79–89.
Henricks, Robert G., trans. *Lao-tzu: Te-tao ching*. New York: Ballantine, 1989.
Ho Peng Yoke. *The Astronomical Chapters of the Chin Shu*. Paris: Mouton & Co., 1966.
Huang Yi-long 黃一農. *Shehui tianwenxue shi* 社會天文學史 (A History of Social Astronomy). Shanghai: Fudan daxue, 2004.
Jiang Xiaoyuan 江曉原. *Tianxue zhenyuan* 天學真原 (The True Origins of Celestial Studies). Shenyang: Liaoning jiaoyu, 1991.
Kalinowski, Marc. *Maître de Huainan. Traité des Figures Célestes*. Paris: Les Belles Lettres, 2022.
———. "The Notion of 'shi' 式 and Some Related Terms in Qin-Han Calendrical Astrology." *Early China* 35–36 (2012–13): 331–60.
———. "La rhétorique oraculaire dans les chroniques anciennes de la Chine. Une étude des discours prédictifs dans le *Zuozhuan*" (The Rhetoric of Oracles in the Ancient Chronicles of China: A Study of Predictive Discourse in the *Zuozhuan*). *Extrême-Orient, Extrême-Occident* 21 (1999): 37–65.
———. "The Use of the Twenty-Eight *xiu* as a Day-Count in Early China." *Chinese Science* 13 (1996): 55–81.
———. "The *Xingde* 刑德 Texts from Mawangdui." Translated by Phyllis Brooks. *Early China* 23–24 (1998–99): 125–202.
Keightley, David N. *Sources of Shang History: The Oracle-Bone Inscriptions of Bronze Age China*, 2nd ed. Berkeley: University of California Press, 1985.
Kern, Martin. "In Praise of Political Legitimacy: The *Miao* and *Jiao* Hymns of the Western Han." *Oriens Extremus* 38, no. 1 (1996): 29–67.
———. "Religious Anxiety and Political Interest in Western Han Omen Interpretation: The Case of the Han Wudi Period (141–87 B.C.)." *Chugoku shigaku* 中国史学 (Studies in Chinese History) 10 (Dec. 2000): 1–31.
Knechtges, David R. "A Journey to Morality." In *Essays in Commemoration of the Golden Jubilee of the Fung Ping Shan Library, 1932–1982*, edited by Chan Ping-leung et al., 162–182. Hong Kong: Fung Ping Shan Library of the University of Hong Kong, 1982.
———., trans. "Rhapsody on Contemplating the Mystery," 105–38. In *Wen Xuan, or Selections of Refined Literature*. Vol. 3. Princeton, NJ: Princeton University Press, 1996.
Korhonen, Kuisma. "Textual Communities: Nancy, Blanchot, Derrida." *Culture Machine* 6 (2006). http://culturemachine.net/community/textual-communities.

Kory, Stephan N. "Omen Watching, Mantic Observation, Aeromancy, and Learning to 'See.'" *East Asian Science, Technology, and Medicine* 50 (2019): 67–132.

Kristeva, Julia. *Desire in Language: A Semiotic Approach to Literature and Art.* Translated by Thomas Gora, Alice Jardine, and Leon S. Roudiez. New York: Columbia University Press, 1980.

Lagerwey, John, and Marc Kalinowski eds. *Early Chinese Religion—Part One: Shang through Han (1250 BC–220 AD)*. Vol. 2. Leiden: Brill, 2009.

Latour, Bruno. *We Have Never Been Modern*. Translated by Catherine Porter. Cambridge, MA: Harvard University Press, 1993.

Legge, James, trans. *The Shoo King*. In *The Chinese Classics with a translation, critical and exegetical notes, prolegomena, and copious indexes*. Vol. 3 in 2 parts. Hong Kong: Hong Kong University Press, 1972.

Lehoux, Daryn. *What Did the Romans Know? An Inquiry into Science and World Making*. Chicago: University of Chicago Press, 2012.

Li Ling 李零 ed., Liu Lexian 劉樂賢 assist ed. *Zhongguo fangshu gaiguan: Zhanxing juan* 中國方術概觀:占星卷 (Outline of Recipes and Techniques in China: Volume on Astro-omenlogy). 2 vols. Beijing: Renmin Zhongguo, 1993.

Li Shisheng 黎石生. "Changsha Mawangdui sanhao muzhu zaiyi" 長沙馬王堆三號墓主再議 (A Re-evaluation of the Identity of the Occupant of Tomb 3 at Mawangdui, Changsha). *Gugong bowuyuan yuankan* 故宮博物院院刊 (Palace Museum Journal), no. 3 (2005): 150–55.

Li Wai-yee. *The Readability of the Past in Early Chinese Historiography*. Cambridge, MA: Harvard University Asia Center, 2007.

Lien, Yeong-Chung E. "Zhang Heng, Eastern Han Polymath, His Life and Works." PhD diss., University of Washington, Seattle, Washington, 2011. UMI (3452743).

Lippiello, Tiziana. *Auspicious Omens and Miracles in Ancient China: Han, Three Kingdoms and Six Dynasties*. Sankt Augustin: Monumenta Serica, 2001.

Liu, Lydia. *Translingual Practice: Literature, National Culture, and Translated Modernity—China, 1900–1937*. Stanford, CA: Stanford University Press, 1995.

Liu Pakyuen (Liao Boyuan) 廖伯源. *Les institutions politiques et la lutte pour le pouvoir au milieu de la dynastie des Han Antérieurs* (Political Institutions and the Struggle for Power in Western Han Dynasty). Paris: Collège de France, 1983.

Liu Qiyu 劉起釪 (1917–2012). *Shangshu xue shi* 尚書學史 (A History of the Study of the *Venerable Documents*). Beijing: Zhonghua, 1989.

Liu Tseng-kuei. "Calendrical Computation Numbers and Han Dynasty Politics: A Study of Gu Yong's Three Troubles Theory." Translated by Luke Habberstad with additional translation of memorials by Michael Nylan. In Nylan and Vankeerberghen, *Chang'an 26 BCE*, 293–322.

Liu Yi et al. "Mysterious Abrupt Carbon-14 Increase in Coral Contributed by a Comet." *Nature* 4: 3728 (Jan. 2014): 1–4.

Lloyd, Geoffrey, and Nathan Sivin. *The Way and the Word: Science and Medicine in Early China and Greece*. New Haven, CT & London: Yale University Press, 2002.

Loewe, Michael. *A Biographic Dictionary of the Qin, Former Han, and Xin Periods: 221 BC–24 AD*. Leiden: Brill, 2000.

———. *Crisis and Conflict in Han China: 104 BC–AD 9*. London: George Allen and Unwin, 1974.

———. *Dong Zhongshu, a 'Confucian' Heritage and the* Chunqiu Fanlu. Leiden: Brill, 2011.

———. *Early Chinese Texts: A Bibliographical Guide*. Berkeley: University of California, 1993.

———. "The Fall of Han." In Twitchett and Loewe, *Ch'in and Han Empires*, 317–76.

———. "The Former Han Dynasty." In Twitchett and Loewe, *Ch'in and Han Empires*, 103–222.

———. "The Han View of Comets," 61–84. In *Divination, Mythology, and Monarchy in Han China*. New York: Cambridge University Press, 1994.

———. "The Tombs Built for Han Chengdi and Migrations of the Population." In Nylan and Vankeerberghen, *Chang'an 26 BCE*, 201–17.

Lu Yang 盧央, *Zhongguo gudai xingzhan xue* 中國古代星占學 (Astro-omenology in Ancient China). Beijing: Zhongguo kexue jishu, 2008.

Lynn, Richard John, trans. *The Classic of Changes: A New Translation of the I Ching as Interpreted by Wang Bi*. New York: Columbia University Press, 1994.

Major, John S. *Heaven and Earth in Early Han Thought: Chapters Three, Four, and Five of the* Huainanzi. Albany: SUNY Press, 1993.

Major, John S., Sarah Queen, Andrew Meyer, and Harold D. Roth, trans. *The Huainanzi: A Guide to the Theory and Practice of Government in Early Han China*. New York: Columbia University Press, 2010.

Mansvelt-Beck, B. J. *The Treatises of Later Han: Their Author, Sources, Contents, and Place in Chinese Historiography*. Leiden: Brill, 1990.

Matsumoto Masaaki 松本雅明 (1912–1993). *Shunjū sengoku ni okeru Shōsho no tenkai: rekishi ishiki no hatten o chūshin ni* 春秋戰國における尚書の展開: 歷史意識の發展を中心に (The Development of the *Venerable Documents* in the Chunqiu and Warring States Periods: Concentrating on the Development of a Consciousness of History). Tōkyō: Kazama Shobō, 1966.

Merton, Robert. "The Puritan Spur to Science" [1938]. In *The Sociology of Science: Theoretical and Empirical Investigations*, 228–53. Chicago: University of Chicago Press, 1973.

Morgan, Daniel P. *Astral Sciences in Early Imperial China: Observation, Sagehood and the Individual*. Cambridge: Cambridge University Press, 2017.

———. "Heavenly Patterns." In Morgan and Chaussende, *Monographs in Tang Official Historiography*, 143–79.

———. "The Planetary Visibility Tables in the Second-Century BC Manuscript Wu Xing Zhan 五星占." *East Asian Science, Technology, and Medicine* 43 (2016): 17–60.

———. "Reflections on Visual and Material Sources for the History of the Exact Sciences in Early Imperial China." *NTM Journal of the History of Science, Technology and Medicine* 28.3 (2020): 325–57.

Morgan, Daniel P., and Damien Chaussende, eds., with the collaboration of Karine Chemla. *Monographs in Tang Official Historiography: Perspectives from the Technical Treatises of the* History of Sui (Sui shu). Cham, Switzerland: Springer, 2019.

Needham, Joseph (in collaboration with Wang Ling). *Mathematics and the Sciences of the Heavens and the Earth*. Vol. 3 of *Science and Civilization in China*. Cambridge: Cambridge University Press, 1959.

Neuhäuser, D. L., R. Neuhäuser, and J. Chapman. "New Sunspots and Aurorae in the Historical Chinese Text Corpus?: Comments on Uncritical Digital Search Applications." *Astronomische Nachrichten* 339, no. 1 (Jan. 2018): 10–29.

Neuhäuser, Ralph, Dagmar L. Neuhäuser, and Thomas Posch. "Terra-Astronomy—Understanding Historical Observations to Study Transient Phenomena." *Proceedings of the International Astronomical Union* 14, no. A30 (2018): 145–47.

Nienhauser, William H. Jr. et al., trans. *The Grand Scribe's Records*. Vol. 1., "The Basic Annals of Pre-Han China." Bloomington & Indianapolis: Indiana University Press, 1994.

Nivison, David, trans. "The Metal-Bound Coffer." In *Sources of Chinese Tradition: From Earliest Times to 1600*, edited by WM. Theodore de Bary and Irene Bloom, 32–35. 2nd ed. Vol. 1. New York: Columbia University Press, 1999.

Nylan, Michael. *The Canon of Supreme Mystery: A Translation with Commentary of the T'ai Hsüan Ching*. Albany: SUNY Press, 1993.

———. "Classics without Canonization: Learning and Authority in Qin and Han." In Lagerwey and Kalinowski, *Early Chinese Religion*, 721–76.

———. *Exemplary Figures: Fayan*. Seattle: University of Washington Press, 2013.

———. "On the *Hanshu* 'Wuxing zhi' 五行志 and Ban Gu's Project." In Csikszentmihalyi and Nylan, *Technical Arts*, 213–79.

———. *The Shifting Center: The Original "Great Plan" and Later Readings*. Nettetal, Germany: Steyler Verlag, 1992.

———. "Sima Qian: A True Historian." *Early China* 23/24 (1998–99): 203–46.

———. "The 'Treatise on the *Wuxing*' (*Wuxing zhi*)." In Morgan and Chaussende, *Monographs in Tang Official Historiography*, 181–233.

———. *Yang Xiong and the Pleasures of Reading and Classical Learning in China*. New Haven, CT: American Oriental Society, 2011.

———. "Yin-yang, Five Phases, and *qi*." In Nylan and Loewe, *China's Early Empires*, 398–413.

Nylan, Michael, and Michael Loewe, eds. *China's Early Empires: A Re-appraisal.* Cambridge: Cambridge University Press, 2010.
Nylan, Michael, and Griet Vankeerberghen, eds. *Chang'an 26 BCE: An Augustan Age in China.* Seattle & London: University of Washington Press, 2014.
Olberding, Garret P. S. *Dubious Facts: The Evidence of Early Chinese Historiography.* Albany: SUNY Press, 2012.
Pankenier, David W. *Astrology and Cosmology in Early China: Conforming Earth to Heaven.* Cambridge: Cambridge University Press, 2013.
———. "The *Huainanzi's* 'Heavenly Patterns' and the *Shiji's* 'Treatise on the Celestial Offices': What's the Difference?" In *The* Huainanzi *and Textual Production in Early China*, edited by Sarah A. Queen and Michael Puett, 199–224. Leiden: Brill, 2014.
Peterson, Willard. "Making Connections: 'Commentary on the Attached Verbalizations' of the *Book of Change.*" *Harvard Journal of Asiatic Studies* 42, no. 1 (Jun. 1982): 67–116.
Puett, Michael. "Sages, Ministers, and Rebels: Narratives from Early China Concerning the Initial Creation of the State." *Harvard Journal of Asiatic Studies* 58, no. 2 (Dec. 1998): 425–79.
Queen, Sarah A. *From Chronicle to Canon: The Hermeneutics of the* Spring and Autumn, *According to Tung Chung-shu.* Cambridge: Cambridge University Press, 1996.
Raphals, Lisa. *Divination and Prediction in Early China and Ancient Greece.* Cambridge: Cambridge University Press, 2013.
Richey, Jeffrey L. "Lost and Found Theories of Law in Early China." *Journal of the Economic and Social History of the Orient* 49, no. 3 (2006): 329–43.
Richter, Matthias L. *The Embodied Text: Establishing Textual Identity in Early Chinese Manuscripts.* Leiden: Brill, 2013.
Riegel, Jeffrey. "Li chi." In Loewe, *Early Chinese Texts*, 293–97.
Robertson, Duncan. "Writing in the Textual Community: Clemence of Barking's Life of St. Catherine." *French Forum* 21, no. 1 (Jan. 1996): 5–28.
Rochberg, Francesca. *The Heavenly Writing: Divination, Horoscopy and Astronomy in Mesopotamian Culture.* Cambridge: Cambridge University Press, 2004.
Roth, Harold D. *The Textual History of the* Huainanzi. Ann Arbor, MI: Association for Asian Studies, 1992.
Schaberg, David. *A Patterned Past: Form and Thought in Early Chinese Historiography.* Cambridge, MA: Harvard University Asia Center, 2001.
Schafer, Edward H. *Pacing the Void: T'ang Approaches to the Stars.* Berkeley: University of California Press, 1977.
Shaughnessy, Edward L. I Ching, *The Classic of Changes: The First English Translation of the Newly Discovered Second-Century B.C. Mawangdui Texts.* New York: Ballantine, 1997.
Sivin, Nathan. "Change and Continuity in Early Cosmology: *The Great Commentary* to *The Book of Changes.*" In *Chūgoku kodai kagakushi ron zokuhen*

中国古代科学史論続編 (Essays on Ancient Chinese Science), edited by Yamada Keiji 山田慶兒, 3–43. Kyoto: Jinbun Kagaku Kenkyūjo, 1989.

———. "Cosmos and Computation in Early Chinese Mathematical Astronomy." *T'oung Pao*, Second Series, 55, Livr. 1/3 (1969): 1–73.

———. *Granting the Seasons: The Chinese Astronomical Reform of 1280, With a Study of its Many Dimensions and a Translation of its Records*. New York: Springer, 2009.

Sleeswyk, André Wegener, and Nathan Sivin. "Dragons and Toads: The Chinese Seismoscope of A.D. 132." *Chinese Science* 6 (Nov. 1983): 1–19.

Smith, Richard J. "Prognostication in Premodern China: Issues of Culture and Class." In *Handbook of Prognostication in China*, Part 1, edited by Michael Lackner and Zhao Lu, 123–73. Leiden: Brill, 2022.

Stellarium. Version 0.22.2. Available for download at www.stellarium.org.

Stock, Brian. *The Implications of Literacy: Written Language and Models of Interpretation in the Eleventh and Twelfth Centuries*. Princeton, NJ: Princeton University Press, 1983.

Sun Xiaochun, and Jacob Kistemaker. *The Chinese Sky During the Han: Constellating Stars and Society*. Leiden: Brill, 1997.

Tao Lei 陶磊. "*Huainanzi Tianwen" yanjiu: cong shushu de jiaodu* 『淮南子·天文』研究——從數術的角度 (Research on the *Huainanzi* "Celestial Patterns"—from a Mathematical Perspective). Jinan: Qi Lu shushe, 2003.

Thote, Alain. "Daybooks in Archaeological Context." In Harper and Kalinowski, *Books of Fate and Popular Culture*, 11–55.

Tomiya Itaru 冨谷至, and Yoshikawa Tadao 吉川忠夫, trans. *Kanjo gogyōshi* 漢書五行志 (The *History of the Han* "Wuxing Treatise"). Tokyo: Heibonsha, 1986.

Tseng, Lillian Lan-ying. *Picturing Heaven in Early China*. Cambridge, MA: Harvard University Press, 2011.

Twitchett, Denis, and Michael Loewe, eds., *The Ch'in and Han Empires*. Vol. 1 of *The Cambridge History of China*. Cambridge: Cambridge University Press, 1986.

Vankeerberghen, Griet. *The Huainanzi and Liu An's Claim to Moral Authority*. Albany: SUNY Press, 2001.

———. "Pining for the West: Chang'an in the Life of Kings and Their Families during Chengdi's Reign." In Nylan and Loewe, *China's Early Empires*, 347–66.

Veeser, H. Aram. Introduction to *The New Historicism*, ix–xvi. Edited by H. Aram Veeser. New York: Routledge, 1989.

Waley, Arthur, trans. *The Analects of Confucius*. New York: Random House, 1989 [1938].

———., trans. *The Way and Its Power: A Study of the Tao Te Ching and Its Place in Chinese Thought*. New York: Grove, 1958 [1934].

Waley, Arthur, trans. and Joseph R. Allen, ed. *The Book of Songs: The Ancient Chinese Classic of Poetry*. New York: Grove Press, 1996.

Wang Aihe. *Cosmology and Political Culture in Early China*. Cambridge: Cambridge University Press, 2000.

Waring, Luke. "Writing and Materiality in the Three Han Dynasty Tombs at Mawangdui." PhD diss., Princeton University, Princeton, New Jersey, 2019. UMI (13864026).

Wen Shiou Tsu. "The Observations of Halley's Comet in Chinese History." *Popular Astronomy* 42 (1934): 191–201.

Wittgenstein, Ludwig. *Philosophical Investigations*. 2nd ed. Translated by G.E.M. Anscombe. 1958. Reprint of English text with index. Oxford: Basil Blackwell, 1986.

Xi Zezong 席澤宗 (1927–2008). "Mawangdui Hanmu boshu zhong de huixingtu" 馬王堆漢墓帛書中的彗星圖 (The Chart on Comets among the Silk Texts from the Han Tombs at Mawangdui). *Wenwu* 2 (1978): 5–9.

Xu Zhentao 徐振韜 (1936–2004). "Cong boshu 'Wuxing zhan' kan 'Xian-Qin hunyi' de chuangzhi" 從帛書《五星占》看"先秦渾儀"的創制 (An Examination of the Invention of the Armillary Sphere in the Pre-Qin Era through the Silk "Five Planets Sign-Readings"). *Kaogu* 考古 (Archaeology) (Feb. 1976): 84, 89–94.

Xu Zhentao, David W. Pankenier, and Jiang Yaotiao. *East Asian Archaeoastronomy: Historical Records of Astronomical Observations of China, Japan, and Korea*. Amsterdam: Gordon and Breach Science Publishers on behalf of the Earth Space Institute, 2000.

Yang Shao-yun. "The Politics of Omenology in Chengdi's Reign." In Nylan and Vankeerberghen, *Chang'an 26 BCE*, 323–46.

Yao Zhenzong 姚振宗 (1843–1906). *Hanshu yiwenzhi tiaoli* 漢書藝文志條理 (Explanation of Items in the *History of the Han* "Bibliographic Treatise"). In *Ershiwu shi yinwen jingji zhi kaobu cuibian* 二十五史藝文經籍志考補萃編 (Assorted Collection of Supplements and Investigations into the Bibliographic Treatises of the *Twenty-Five Standard Histories*), edited by Wang Chenglüe 王承略 and Liu Xinming 劉心明. Beijing: Qinghua daxue, 2011–2012.

Yasui Kōzan 安居香山 (1921–1989) and Nakamura Shōhachi 中村璋八 (1926–1993) eds. *Jōshū isho shūsei* 重修緯書集成 (Revised Collection of Apocryphal Texts). 6 vols. Tokyo: Meitoku, 1971–1992.

Yates, Robin D. S. *Five Lost Classics: Tao, Huang-Lao, and Yin-yang in Han China*. New York: Ballantine, 1997.

———. "The History of Military Divination in China." *East Asian Science, Technology, and Medicine* 24 (2005): 15–43.

Zhang Xincheng 張心澂 (1887–1973). *Weishu tongkao* 偽書通考 (Comprehensive Investigation into Forged Books). Shanghai: Shangwu, 1939.

Zhou Yiqun. "Helen and Chinese Femmes Fatales." In *Ancient Greece and China Compared*, edited by G.E.R. Lloyd and Jingyi Jenny Zhao, in collaboration with Qiaosheng Dong, 234–56. Cambridge: Cambridge University Press, 2018.

Zotti, Georg et al. "The Simulated Sky: Stellarium for Cultural Astronomy Research." *Journal of Skyscape Archaeology* 6, no. 2 (2020): 221–58.

Zürn, Tobias Benedikt. "The Han Imaginaire of Writing as Weaving: Intertextuality and the *Huainanzi*'s Self-Fashioning as an Embodiment of the Way." *Journal of Asian Studies* 79, no. 2 (2020): 367–402.

———. "The Practice of Erasing Traces in the *Huainanzi*." In *The Craft of Oblivion: Forgetting and Memory in Ancient China*, edited by Albert Galvany, 181–223. Albany: SUNY Press, 2023.

Index

*Dates for historical figures are given on the first occurrence in the text.
Stars and constellations appear under *asterisms* and *lunar lodges*.
For entries on terms in Chinese, English translations in some instances appear on the following page.

Allan, Sarah, 191n31, 210n76
Analects (Lunyu 論語), 79, 176; quotations from, 148, 220n52, 224n28
Annals (Chunqiu 春秋), 180; and "Celestial Patterns," 98–99, 174; and memorials, 105–6, 112, 116, 145–46, 149, 175, 177, 224n32, 227n65; and "Wuxing Treatise," 17, 80–88, 100, 174. See also Five Classics, *Gongyang*; Kongzi; *Zuozhuan*
Annals of Lü Buwei. See *Lüshi chunqiu*
apocrypha, 14, 212n8, 218n24, 219n38; in Xiang Kai's memorial, 117–19; Zhang Heng, views on, 162, 228n82
"Appended Statements" (Xici 繫辭): "Celestial Offices," citation of, 64, 208n50; in Mawangdui mss., 40–41, 44, 47, 201n94, 202n95; in memorials, 17, 106–10, 120–21, 138, 175, 177, 227n70; Peterson's view of, 202n99; *tianwen*, description of, 6; "Wuxing Treatise," citation of, 214n17
asterisms: Assistant Conductors 攝提格, 52, 158, 204n6; Grand Stairway 泰階, 185–86, 233n5; Grand Tenuity 太微, 100, 136, 222n90; Great Horn 大角, 88, 214n25, as Big Horn, 30; Great Fire 大火, 4, as Great Star 大辰, 87; Jade Crossbar 玉衡, 185–86, 233n6; Northern Dipper 北斗, 14, 26, 51–54, 87, 225n42; Oxherd 牛郎, 99; Purple Palace 紫宮, 158, 227n67. See also lunar lodges
Aveni, Anthony, 10

Bactria, 50
Ban Biao 班彪, 215n43
Ban Gu 班固: Dong Zhongshu, biography of, 145; *Hanshu* treatises, 79–81, 90, 101, 214n17; Liu Xiang, biography of, 227n63; Marquis of Dai, third, dating of, 193n6. See also *Hanshu*
Ban Zhao 班昭, 81, 90, 97, 101, 176

· 249 ·

Bao Si 褒姒, 75–76
"Basic Annals" (Benji 本紀). See under Sima Qian
Bei Gong 北宮 (Northern Palace), 32
Bi Gan 比干, 70
bian 變 (aberrate/aberration; alternate/alternation), 181; as aberration, 62, 93, 98, 108, 116, 119, 124, 125, 129, 130, 136, 137, 156, 158, 162, 169; in bian hu 變忽 (alter and disorder), 164; as dynastic change, 155; as personal change, 124, 165; as sign, 99; in shibian 時變 (change of the seasons), 6, 7, 66; in titles of texts, 186, 188n9; as transformation, 57
Bian Shao 邊韶, 219n38
Bie lu 別錄 (Catalogued Records), 187n4
Boyang Fu 伯陽甫, 75
Brown, David, 31, 198n59

Cai Yong 蔡邕, 136–38, 219n38
calendrics. See li
"Canon of Yao" (Yao dian 堯典): historicity, debate on, 191n31; and Shiji correspondent, 67–68, 76, 209n68, 210n84, 218n33
Cao Jie 曹節, 130
"Celestial Offices, Treatise on" (Tianguan shu 天官書), 6–7, 50–57, 63–66, 76–78, 205nn10–11, 208n55, 232n3. See also "Celestial Patterns, Treatise on"
"Celestial Patterns, Treatise on" (Tianwen zhi 天文志): "Celestial Offices" and, 6–7, 66, 81, 90–101, 216n54; compilation of, 81, 90, 176, 229n95; Liu Xiang, comparison with language used by, 227n64; Mars guarding Heart, record of, 132, 134; tu qi 土氣, use of phrase, 216n53; and "Wuxing Treatise," 17, 82, 90, 96–101, 174; Xingzhuan 星傳, citations of, 214n21
Changes (Yijing 易經), 2, 180; apocrypha, related to, 117; in "Celestial Patterns," 95–96; creation of hexagrams in, 69, 82; in Huainanzi, 77; Jing Fang Tradition of, 213n11; Ma Rong's commentary on, 176; in Mawangdui mss., 23, 37, 40–48, 174, 202nn95–96; in memorials, 105, 128–29, 158–59; tianwen, descriptions of, 6–7; in "Wuxing Treatise," 82–83, 86, Yang Xiong's emulation of, 79. See also "Appended Statements"; bian; Eight Trigrams; Sixty-four Hexagrams
Chao Hong 晁閎, 124
Chen Bao 陳褒, 116–17, 135, 137
Chen Fan 陳蕃, 119, 130–31
Chen Meidong 陳美東, 13
Chen Zhong 陳忠, 116–17, 135–38
Cheng Jin 成瑨, 117
Chu 楚, 31–32, 65, 88, 136, 199n61, 207n45, 209n57, 221n73. See also Chuci
Chuci 楚辭, 151, 199n74, 231n6
Chunqiu 春秋. See Annals
Chunqiu period, 2, 53, 65, 80, 82, 87–88, 94, 98, 175, 212n5
Classic of Filial Piety (Xiaojing 孝經), 117–19
clouds, strange or oddly-shaped, 1, 13, 23, 54, 121; in court liturgy, 152; and qi, 4, 53, 59–60, 86, 92; texts for reading, 7, 26, 31–35, 47, 185–86; Yao, in comparison with, 67

Index | 251

comets: Banner of Chiyou, 89, 215n33; in "Celestial Patterns," 98–99; in Mawangdui mss., 32–33; in memorials, 108, 119, 121, 133; radioisotopes, as cause of spike in, 192n45; in *Shiji*, 63–66, 92, 208n55; in titles of texts, 185–86; in "Wuxing Treatise," 82, 87–90, 214n22; as *xiao* 181; Yanzi and Duke Jing, in debate between, 3–5; as *zaiyi*, 180; in Zheng, 4. See also Halley's Comet
Confucius. See Kongzi
Constant Models (Jing fa 經法), 23, 37–40, 44, 46–47, 57, 174, 201nn82–83
Csikszentmihalyi, Mark, ix, xi, 203n117
Cullen, Christopher, 25, 30–31, 189n14, 197n28, 209n72

Da Yin 大音. See Great Sound
Dai, Marquises of, 22–23. See also Li Cang
Daji 妲己, 73
Dan Zhu 丹朱, 68, 76
Daodejing 道德經 (Classic of the Way and Its Power), 2, 23, 37, 77, 201n83
Dazzling Deluder 熒惑. See Mars
de Crespigny, Rafe, 117
de 德: as favor or Favor, 29, 34, 38, 39, 51–52, 56–57, 146, 148, 184, 196n22, 205n14; as virtue or character, 27, 28, 36–37, 44–46, 48, 72, 91, 115, 118, 123, 125, 126, 129, 130, 137, 142, 148–51, 155–61, 182, 184, 203n116, 205n14, 210n76, 211n92, 227n55; "Three virtues," 111, 112, 116; *xiong de* 凶德, 184, 232n18. See also "Harshness and Favor"

Di Kongjia 帝孔甲, 71
Di Zhongkang 帝中康, 71
Ding Hong 丁鴻, 106–7, 120
"Discourse on the Wuxing Tradition of the 'Great Plan'" (Hongfan Wuxing zhuan lun 洪範五行傳論), 79, 212n4
Documents (Shangshu 尚書): "Basic Annals," in comparison with, 67, 71–72, 74, 77, 207n46, 210n86, 210nn89–90; in "Celestial Patterns," 95–96; in Five Classics 2, 43, 105, 119, 145, 175; "Luo gao" 洛誥, 228n71; Ma Rong, commentary on, 176; Ouyang interpretation of, 107, 112, 217n4; "three-fold examination" in, 114. See also "Earl of the West Conquers Li"; Fu Sheng; "Great Oath"; "Great Plan"; "High Ancestor's Day of Sacrifice"; "Metal-bound Coffer"; "Oath at Gan"; Tang
Donald, Merlin, 10
Dong Xian 董賢, 100
Dong Zhongshu 董仲舒: on administration and seasons, 184; Confucianism, debates over role in establishing, 201n84, 223n10; dates, 213n10; Emperor Wu, memorials to, 143–53, 155, 170, 175, 223n13, 223n19, 230n4; exile to Jiaoxi, 224n32; imperial family, attacks on collateral branches of, 206n18; in "Wuxing Treatise," 82, 87, 213n11; *zaiyi*, use of phrase, 231n6
Dongfang Shuo 東方朔, 233n5
Dou Miao 竇妙, 119
Dou Rong 竇融, 106, 113
Dou Wu 竇武, 119, 130–31
Dou Xian 竇憲, 107

Dowager Deng 鄧, 107, 116
Dowager Ding 丁, 128
Dowager Dou 竇 (d. 135 BCE), 144
Dowager Dou 竇 (d. 97 CE). *See* Empress Dou
Dowager Empress Fu 傅, 125–26
Dowager Wang 王, 153, 225n45
Du Ye 杜鄴, 128–31, 158
Duke Huan 桓 of Qi, 108, 209n57, 218n33
Duke Jing 景 of Qi, 3–4, 36–37, 47, 175
Duke Jing 景 of Song, 62, 122, 131–39, 221n73
Duke of Shao 召, Co-Regent (r. 841–828), 75, 211n106
Duke of Zhou 周 (11th cent. BCE), 122, 127–31, 138, 148, 221nn70–71, 228n71
Duke of Zhou 周, Co-Regent (r. 841–828), 75, 211n106
Duke Wen 文 of Jin, 207n45, 209n57, 218n33
Duke Xiang 襄 of Lu, 157, 227n65
Duke Xiang 襄 of Song, 65, 209n57

"Earl of the West Conquers Li" (Xibo kan Li 西伯戡黎), 73–74
Earl of Wind (Feng Bo 風伯), 34, 199nn73–74
earthquake, 1, 3, 177; of 133 CE, memorials on, 17, 143, 160–71, 229n95; in additional memorials, 116, 119, 120, 129, 135, 138; in "Basic Annals," 75–76, in "Celestial Offices," 92; in *Huainanzi*, 60; in Mawangdui mss., 28–31; Zhang Heng's detection of, 162, 181, 229n84
eclipse, 1, 10, 12, 13, 180–81, 186; in "Celestial Offices," 65–66; in "Celestial Patterns," 91; in *Huainanzi*, 59, 61; in Mawangdui mss., 23, 26, 29; in memorials, 106–7, 113–20, 125–29, 131, 144, 153, 155, 157–58, 163, 168–69, 177, 220n54, 223n9, 227n65; in "Wuxing Treatise," 82; in *Zuozhuan*, 199n61
Eight Trigrams, 41, 43, 82
Eliade, Mircea, 10
Emperor Ai 哀, 90, 100, 125, 160, 228n82
Emperor An 安, 107, 116–17, 120, 135
Emperor Cheng 成: bibliographic project, 2, 79, 187n4; and comet of 32 BCE, 90; infant children, death of, 153, 225n46; Mars guarding Heart, 133–37, 222n87; nomination of heir, 215n37; omen of flock of pheasants, 123–25, 220n50; phases of omen use, 225n47; Wang clan, rise in power, during reign of, 99. *See also* Gu Yong, Halley's Comet, Liu Xiang, Zhai Fangjin
Emperor Guangwu 光武, 99, 109, 112–16
Emperor He 和, 106, 109
Emperor Huan 桓, 117, 119
Emperor Jing 景, 50, 89, 158, 227n65
Emperor Ling 靈, 119–20, 131, 137
Emperor Ping 平, 228n82
Emperor Shun 順, 107, 117; and earthquake of 133 CE, 160–63, 167, 171
Emperor Wen 文, 89, 109, 193n6
Emperor Wu 武, 49–50, 79, 89, 98, 208n55; absence of auspicious signs during reign of, 17, 143–55 passim, 170, 175, 223n8
Emperor Xuan 宣, 89–90, 157, 223n9
Emperor Yuan 元, 98, 143, 175

Index | 253

Emperor Zhang 章, 106
Empress Chen 陳, 89, 214n31
Empress Dou 竇, 106
Empress Lü 呂, 66, 128, 193n5
Empress Wei 衛, 89, 214n31
Empress Yan 閻, 107, 217n10
"Encountering Sorrow." See "Li sao"
"Er san zi wen" 二三子問. See "Inquiries of Two or Three Disciples"
"Essentials" (Yao 要), 40, 43–44, 48, 177
eunuchs, 117–21, 217n17

fa 罰 (punishment; penalty), 182–83; in *Hanshu* treatises, 85, 96; in memorials, 112, 165; in *tian fa* 天罰 (Punishment of Heaven), 63, 67, 70–72
famine: earthquake of 133 CE and, 160, 164, 166, 168; under Emperor Cheng, 133, 156; in Mawangdui mss., 23, 29, 33, 37
Fei Li 賁麗, 133, 135, 137
Feng Bo 風伯. See Earl of Wind
Feng Long 豐隆, 34
Feng Shi 馮時, 13, 197n26
field allocations (fenye 分野), 4, 53, 91–92
Five Classics, 1–2, 16–17, 77, 79, 105, 114, 131, 174, 187n4. See also *Annals*; *Changes*; *Documents*; *Odes*; *Rites*
Five Kinds of Action (Wuxing 五行), 23, 45–48, 182
Five Planets, 7, 74, 209n74; correspondences to, 28, 37, 54–56, 96, 121, 205n13; as Five Spirits, 58; in *Hanshu* bibliographic treatise, 7, 185–86, 233n7; losing courses, 61; in memorials, 155; in technical treatises, 51–53, 61, 64, 66, 90–94. See also Jupiter; Mars; Mercury; Saturn; Venus
"Five Planets Sign-Readings" (Wuxing zhan 五星戰), 25–31, 36–37, 39, 48, 54–55, 196n21, 197n30, 205n13
floods, 1, 3–4, 180; in Mawangdui mss., 27–32, 37; in memorials, 109, 116–17, 120, 154, 165–69; in treatises, 58–62, 92; under Yao, 145, 225n39
fu 符 (tally), 93, 94, 145, 147, 180
fu rui 符瑞 (auspicious tallies and emblems), 80, 154, 180
Fu Sheng 伏生, 209n74, 214n15

"Gan shi" 甘誓. See "Oath at Gan"
Gao You 高誘, 189n18
"Gaozong rong ri" 高宗肜日. See "High Ancestor's Day of Sacrifice"
Gongsun Hong 公孫弘: Dong Zhongshu, involvement in exile of, 224n32; late career, 224n34; Liu An, investigation of, 50; memorials, 143–45, 149–53, 170, 175, 224n33
Gongyang zhuan 公羊傳 (Gongyang Tradition), 80, 146, 180–81, 212n5; as a broader practice, 175, 223n10, 231n8
Goodman, Howard, 15
"Great Oath" (Tai shi 太誓), 152, 210n81, 224n26; King Wu's vow to carry out Punishment of Heaven in, 70, 210n81
"Great Plan" (Hong fan 洪/鴻範), 17, 208n49; and "Celestial Patterns," 93, 95, 216n48; Five Proofs of, 226n51; *huangji* 皇極 in, 84–85, 111, 213n13; in memorials, 106, 110–21, 125, 135, 138, 154, 166, 174–75, 177, 219n44; Nine

"Great Plan" (Hong fan 洪/鴻範) (*continued*)
Divisions of, 82–83, 110–11, 113, 120, 181; *wuxing* in, 232n14; and "Wuxing Treatise," 79–85, 100, 182, 212n4. See also *wuxing* and *wushi*
Great Sound (Da Yin 大音), 34, 36, 199n73
Great White 太白. See Venus
Gu Tiefu 顧鐵符, 24
Gu Yong 谷永, 90, 143, 153–61, 168, 170, 181, 225n44, 226n49, 226n51, 226n55
Guan Shuxian 管叔鮮, 127, 220n63
Guan Zhong 管仲, 218n33
guest stars, 7, 108, 186
Gugong Danfu 古公亶父, 212n110
Guliang 穀梁 Tradition, 112
Guo Ji 郭伋, 115

Halley's Comet (12 BCE), 17, 90, 98, 128, 143, 153–60, 215n38, 227n66
halos, 1, 13; in *Hanshu* bibliographic treatise, 186; in Mawangdui mss., 26, 32–36; in *Shiji*, 54
"Harshness and Favor" (Xingde 刑德), 25–26, 33–37, 39, 54, 196n22, 196n24
"Hong fan" 洪/鴻範. See "Great Plan"
"Hongfan Wuxing zhuan lun" 洪範五行傳論. See "Discourse on the Wuxing Tradition of the 'Great Plan'"
Hou Han shu 後漢書 (History of the Later Han), 193n49, 219n38
Hanshu 漢書 (History of the Han), 6–8, 14, 53, 80, 133, 145, 159, 162, 174, 185–86. See also "Celestial Patterns, Treatise on"; "Wuxing Treatise"

"High Ancestor's Day of Sacrifice" (Gaozong rong ri 高宗肜日): in memorials, 122–26, 129, 152, 158; *Shiji* correspondent, 72, 211n102
Hu Yan 狐偃 (aka Jiu Fan 咎犯), 207n45
hua 化 (transform/transformation), 6, 61, 108, 119, 165, 181, 230n102
Huainanzi 淮南子: Dong Zhongshu and, 151, 206n18; heaven and human realm, interactions between, 60–63; Ma Rong, commentary on, 176; Mars guarding Heart, 62, 132, 221n73; *Odes*, citations of, 77, 207nn39–40; sage in, 204n1; and *Shiji*, 15–16, 48, 49–50, 63, 70, 77–78, 174, 177; "Summary of Essentials" (Yao lue 要略), 206n20, 206n22; "Tianwen xun" 天文訓, 6, 50–60, 204n5, 205n11, 207n33; weaving metaphor in, 206n17; –*xun* 訓, addition to chapter titles, 189n18. See also Liu An
Huang Yi-long 黃一農, 134, 221n73
huangji 皇極. See under "Great Plan"
huo 禍/旤 (calamity), 4, 182; as disaster, 112; and good fortune, 80, 93, 109, 124, 126, 163; minuscule beginnings of, 155, 163; Wang Mang usurpation as, 99; in "Wuxing Treatise," 83–85, 183
Huo Guang 霍光, 89
Huo Qubing 霍去病, 215n34

"Inquiries of Two or Three Disciples" (Er san zi wen 二三子問), 40–44, 48

ji 極 (extreme manifestation), 85, 182–83; in Six Extremes (*liu ji* 六極), 111–13, 125

Jia Kui 賈逵, 219n38
Jiang Xiaoyuan 江曉原, 5, 12, 216n54
"Jiaosi ge" 郊祀歌. See "Songs for the Suburban Sacrifices"
Jie 桀, 32, 71–74, 151, 210n90
Jin Zhuo 晉灼, 205n14
"Jinteng" 金縢. See "Metal-bound Coffer"
Jing fa 經法. See *Constant Models*
Jing Fang 京房 the Younger, 117, 213n11
jiu 咎: as blame, 135; as proof of blame (*jiu zheng* 咎徵), 125, 160, 162, 181; as sign of blame, 74, 85, 120, 160, 182–83
Jizi 箕子, 70, 80, 110
Jupiter: in "Great Plan," 110–11, 208n49; in Mawangdui mss., 25–30, 36–37; in memorials, 118, 121; retrograde movement of, in 1048 BCE, 210n79; in treatises, 54–55, 91, 98, 100 (as Year Star). See also Five Planets

Kalinowski, Marc, 10–11, 192n47, 200n81, 204n5, 223n7
King Cheng 成, 75, 122, 126–31, 138, 147, 152, 157–58, 163, 221n70, 228n71
King He 賀 of Liang, 98
King Jia 甲, 73
King Kang 康, 75
King Li 厲, 74–75, 148
King of Jiaoxi 膠西, 224n32
King Pan'geng 盤庚, 72
King Tai 大王, 127. See also as Gugong Danfu
King Taiwu 太戊, 221n92
King Wen 文, 46, 127, 128, 147, 226n55; in "Basic Annals," 67, 69, 73, 75–76; and the *Changes*, 43, 82

King Wu 武, 62–63, 70, 74–76, 80, 109, 110, 127, 131, 147–48, 152, 207n47, 210n79, 220n63, 224n27
King Wuding 武丁, 67, 72, 122–26, 128–29, 131, 138, 157, 211n102, 219n49
King Wuyi 武乙, 73
King Xin 欨 of Yelang 夜郎, 98
King Xuan 宣, 67, 75, 123
King You 幽, 75–76, 148
King Zhow 紂, 63, 70, 73–75, 110, 227n55
King Zhuang 莊 of Chu, 136, 209n57
Kistemaker, Jacob, 12
Kong Guang 孔光, 125–26, 128, 220n61
Kongzi 孔子, 42–44, 163; and *Annals*, 80, 82, 212n5. See also *Analects*
Kristeva, Julia, 9

Latour, Bruno, 173, 188n11
Lehoux, Daryn, 11
Lei Gong 雷公. See Lord Thunder
Lei Pi 雷被, 50
Lesser White 小白. See Mercury
li 曆: as calendar or calendrical, 6, 25, 51–54, 153, 184, 191n31, 204n7, 208n49, 219n38, 228n82; as mathematical astronomy, 5–6, 12, 189n14, 207n43
Li Cang 利蒼, 22–23, 174, 193n2, 193n6
Li Chunfeng 李淳風, 8
Li Gu 李固, 143, 161, 167–71
Li Guang 李廣, 208n55
Li He 李郃, 107–8, 217n10
Li Ling 李零, 12
Li Ling 李陵, 50
"Li sao" 離騷經 (Encountering Sorrow), 2, 176, 199n74
Li Tong 李通, 115

256 | Index

Li Wai-yee, 10–11, 142
Li Xi 李翕, 10
Li Xi 利豨, 22, 194n6
Li Xian 李賢, 222n86
Li Xun 李尋, 133–34, 137
Li Zhucang 黎朱蒼. See as Li Cang
Liang Ji 梁冀, 160
Liang Na 梁妠, 109, 160
lightning: in *Huainanzi*, 60; in liturgy, 152; in Mawangdui mss., 34, 35, 41, 199n71; in memorials, 130, 161–62; in "Metal-bound Coffer," 127
Liji 禮記. See Rites: *Record of Rites*
Liu An 劉安, 15, 49–50, 88–89, 204n3, 225n34
Liu Bang 劉邦, 22, 66, 87–88, 98, 193n1, 193nn4–5, 214n29
Liu Hong 劉洪, 12
Liu Qian 劉遷, 50
Liu Tseng-kuei 劉增貴, 153
Liu Xiang 劉向, 14, 176, 219n49, 230n3; bibliographic project, work on, 79, 174, 187n4, 188n6; Halley's Comet, memorial on, 128, 143, 153, 157–61, 170, 175, 225n44, 227nn63–66, 227n70, 228n71; poetic works of, 231n6; and "Wuxing Treatise," 79–82, 87–90; 100, 212n4, 213n11
Liu Xin 劉歆, 79–82, 87, 100, 114, 174, 187n4
Liu Zhi 劉質/瓆, 117
Liu Zhiji 劉知幾, 212n4
Lloyd, Geoffrey, 11, 173
Lord Thunder (Lei Gong 雷公), 34, 199nn73–74
Lü Shangfu 呂尚父, 70
Lu Yang 盧央, 12
Lu Zhi 盧植, 119–21, 189n18, 219n44

Lun heng 論衡 (Balanced Discourses). See Wang Chong
lunar lodges, 7, 14, 87; in "Celestial Offices," 53–55, 62, 64; day count, use as, 15, 192n47; in *Hanshu* bibliographic treatise titles, 185–86; in *Huainanzi* 52, 62; in Mawangdui mss., 29–31; named instances of, Axletree 軫, 95, 216n46; Barrens 虛, 5, 188n10; Basket 箕, 95, 216n46; Cart Ghost 輿鬼, 98, 216n55; Chamber 房, 29, 198n44; Dipper 斗, 56, 205n14; Eastern Well 東井, 66, 158, 181, 209n62; Heart 心, 29, 118, 121, 132–38, 198n44, 221n73; House 室, 90, 118, 215n36; Net 畢, 95, 216n46
Lunyu 論語. See *Analects*
Lüshi chunqiu 呂氏春秋 (Annals of Lü Buwei), 132

Ma Guohan 馬國漢, 233n3
Ma Rong 馬融, 14, 68, 176, 209n74, 230n3; "Celestial Patterns," work on, 81; earthquake of 133 CE, memorial on, 143, 161, 163–66, 170–71, 229n95; as Lu Zhi's teacher, 119; poetic work of, 231n6
Ma Xu 馬續, 81
Ma Yuan 馬援, 106, 113, 115
Marquis of Shen 申侯, 75–76
Mars: in "Celestial Patterns," 96–98; Duty of Vision, correspondence to, 216n50; in Grand Tenuity, 222n90; guarding Heart and Duke Jing's response, 62, 132–38, 207n43, 221n73; in Mawangdui mss., 26–32; and "Monthly Ordinances," 216n52; retrograde movements,

early knowledge of, 209n66. *See also* Five Planets
Master of Rain (Yu Shi 雨師), 34, 199nn73–74
Master Yan. *See* Yanzi
mathematical astronomy. *See* li 曆
Mawangdui, 14, 16, 21–48, 49, 54–55, 57, 77, 106, 174, 177, 182, 193–204 *passim*
Meng Kang 孟康, 233n3, 233n5
Mercury, 28–30, 57, 197n30. *See also* Five Planets
"Metal-bound Coffer" (Jinteng 金 縢), 126–31; citation by Dong Zhongshu, 145, 147; citation by Liu Xiang, 159
meteor shower, 155
ming 命 (charge, mandate; order, command; life, lifespan), 183–84; as charge or mandate, 67–71, 74, 94, 123, 126, 132, 145, 147, 155, 180; as command or order, 65, 128, 211n98; as naming, 73; in titles of texts, 118, 188n9, 215n43
Mingtang 明堂 (Hall of Illumination), 118, 216n52
"Miscellaneous Sign-Readings on Astronomical and Meteorological Phenomena" (Tianwen qixiang zazhan 天文氣象雜占), 25–26, 31–39, 48, 197n27, 199n61, 209n60
"Monthly Ordinances" (Yue ling 月 令), 93, 96, 124, 216n52
Morgan, Daniel, 8, 12, 25, 30
"Muhe" 繆和, 40, 42–44, 48, 202n95

nie 孽 (bane), 83–85, 112, 182–83
Nine Palaces 九宮: diagram, 26, 34, 196n22; Zhang Heng, expertise in, 228n82

"Nine Songs" (Jiu ge 九歌), 151
novae, 54, 92, 232n3. *See also* guest stars
Nylan, Michael, ix, 80, 212n6, 213n13, 226n48, 232n14

"Oath at Gan" (Gan shi 甘誓), 71, 207n46
"Oath of Tang" (Tang shi 湯誓). *See* Tang: in *Shiji*
Oedipus, 231n7
Odes (Shijing 詩經), 2, 43, 176, 216n47; in "Celestial Patterns," 95–96; in *Huainanzi*, 61, 77, 207nn39–40; in memorials, 105, 145–47, 155, 175, 224n23, 224n25, 226n55. *See also* Five Classics

Pankenier, David, 10, 54, 190n31, 210n79, 217n62, 233n3
planets. *See* Five Planets; Jupiter; Mars; Mercury; Saturn; Venus
Proscribed Faction (dang gu 黨錮), 119–20, 130
Punishment of Heaven. *See under fa* 罰

qi 氣 3; in "Celestial Patterns," 92, 96–97, 216n53; in *Hanshu* bibliographic treatise, 7, 185–86; in *Huainanzi*, 51, 57–61; in *Lun heng*, 4, 188n9; in Mawangdui mss., 25–26, 33–35, 196n22, 196n24; in memorials, 109, 112–13, 116, 124, 148–50, 154; in *Shiji*, 53, 56, 64–66, 75, 92, 208n55, 209n60; in "Wuxing Treatise," 86, 88, 113; and *zhan hou* 占候, 200n81. *See also yin* and *yang*
Qi lue 七略 (Seven Summaries), 187n4

Qi 啟, 71, 207n46
Qi 齊, 5, 65, 87. *See also* Duke Jing of Qi; Duke Huan of Qi
Qin 秦: dynasty, 65–66, 80, 85, 87, 94, 100, 175; model of governance, 79; as predynastic hegemon, 65; unification and collapse, 22, 30, 33, 53, 88, 173, 193n3, 214nn27–28. *See also* Duke Mu of Qin; Qin Shihuang
Qin Shihuang 秦始皇, 26, 65, 157; as First Emperor, 22
Quelling Star 填星. *See* Saturn

rain: of blood, 183; in "Celestial Patterns," 95–96; in edict of Emperor Wu, 145; in "Great Plan," 110, 154, 181, 226n51; in *Hanshu* bibliographic treatise, 7, 185–86; in *Huainanzi*, 58–62; in liturgy, 152; in Mawangdui mss., 26, 34–36, 41, 42, 196n22; in memorials, 109, 116, 120, 135, 138, 149, 164; and mistreatment of Duke of Zhou, 128, 221n70; in "Monthly Ordinances," 216n52; summoning of, 4, 160; in "Wuxing Treatise," 83; in *Yanzi chunqiu*, 3. *See also* Master of Rain
rainbows, 1, 13; in "Celestial Offices," 54; in *Hanshu* bibliographic treatise, 7, 186; in Mawangdui mss., 23, 26, 32–33
Ren shi 任氏, 32
Richey, Jeffrey, 38
rites, 27–28, 37, 43–44, 96, 120, 126, 128, 135, 216n51, 220n61
Rites (Classics), 2, 119; *Record of Rites* (Liji 禮記), 93, 96, 176, 216n50; *Rites of Zhou* (Zhou li 周禮), 166, 176, 229n102. *See also* "Monthly Ordinances"

rui 瑞 (auspicious emblem), 143, 147, 180. *See also furui*

Saturn, 26–28, 30, 55, 57, 98, 133, 197n30, 197n33, 205nn13–14, 216n53. *See also* Five Planets
Schaberg, David, 10–11, 142
Schafer, Edward, 9, 180
Seven Kingdoms Rebellion, 50, 66, 88
Shangshu 尚書. *See Documents*
sheng 眚 (pestilence), 83, 160, 182–83
Shiji 史記. *See* Sima Qian; "Celestial Offices, Treatise on"
Shijing 詩經. *See Odes*
Shun 舜: in memorials, 114, 118, 208n49; in *Shiji* and "Canon of Yao," 67–69, 71, 73, 76. *See also xuanji yuheng*
Sima Biao 司馬彪, 193n49, 218n21
Sima Qian 司馬遷, 15, 49; "Basic Annals" (Benji 本紀) 67–76; "Letter to Ren'an," 204n4; Marquis of Dai, third, dating of, 193n6; Mars, on motion of, 207n43, 209n66; as Senior Archivist, 161; *Shiji*, reasons for compiling, 50, 208n54; *Shiji*, structure of, 208n48. *See also* "Celestial Offices, Treatise on"; *Documents*: "Basic Annals," in comparison with; Sima Tan
Sima Tan 司馬談, 64, 189n19, 207n43
Sima Xiangru 司馬相如, 200n74
Sivin, Nathan, ix, 5, 11, 12, 173
Six Disruptions (*liu li* 六沴), 112–13, 218n22
Sixty-four Hexagrams, 43, 69, 82, 180
Song E 宋娥, 160–61, 167, 169
"Songs for the Suburban Sacrifices" (Jiaosi ge 郊祀歌), 151–52

Sun Xiaochun, 12

Taibo 太白. *See* Venus
Taijia 太甲, 72
"Tai shi" 太誓. *See* "Great Oath"
Tang 湯: in *Shiji* 71–72, 207n46, 210n89; in memorials: 145, 151; as one of five sage-kings, 208n49
Tang Du 唐都, 64–65, 208n55
"Tang shi" 湯誓 (Speech of Tang). *See* Tang, in *Shiji*
"Tianguan shu" 天官書. *See* "Celestial Offices, Treatise on"
tianwen 天文: Babylonian "heavenly writing," in comparison with, 189n15; in *Changes*, 6, 44, 47; versus *li* or astronomy, 5–6, 12–13; as technical bibliographic category, 7–8, 10, 16, 182, 185–86. *See also* "Celestial Patterns, Treatise on"; *Huainanzi*: "Tianwen xun"
"Tianwen qixiang zazhan 天文氣象雜占. *See* "Miscellaneous Sign-Readings on Astronomical and Meteorological Phenomena"
"Tianwen xun" 天文訓. *See under Huainanzi*
"Tianwen zhi" 天文志. *See* "Celestial Patterns, Treatise on"
Tseng, Lillian, 9–10, 190n30

Venus, 57; in Mawangdui mss., 25–31, 197n30; in memorials, 118, 121, 133, 137, 222n90. *See also* Five Planets

Wang Chong 王充, 4, 188n9, 221n70
Wang Ji 王季, 127
Wang Mang 王莽, 80–82, 87–90, 99–101, 160, 216n58
Wang Shuo 王朔, 64–65, 208n55

Wang Yin 王因, 123–26, 219n50, 225n45
Watch Star 晨星. *See* Mercury
Wei Qing 衛青, 215n34
Wei Xian 魏鮮, 64–65, 208n55
Wei Zifu 衛子夫. *See* Empress Wei
Weizi 微子, 70
Wittgenstein, Ludwig, 9
Wu Rui 吳芮, 193n1
wushi 五事 (Five Duties), 83–86, 93, 96–97, 100–101, 110–11, 113, 116, 120, 182
wuxing 五行, 110–12, 182; earliest sense of, 232n14; in *Huainanzi*, 51; later treatises on, 213n8; in *Shiji*, 64, 71, 210n82; in "Wuxing Treatise," 80, 83, 85, 86, 100; and *yinyang* dynamics, 119, 121, 178, 179, 232n13. *See also* "Wuxing Treatise"; *Five Kinds of Action*
"Wuxing Treatise" (Wuxing zhi 五行志), 17, 79–101 passim, 110, 112–13, 120, 174–75, 182–83, 213n11, 214n21, 219n38

Xi 羲 and He 和, 68, 71, 210n84
Xi Hu 郄縠, 218n33
Xi Zezong 席澤宗, 24
Xia Heliang 夏賀良, 99
Xiahou Shichang 夏侯始昌, 214n15
xiang 祥: as portents or omens, 3, 83–85, 182–83; as auspicious or auspicious sign, 61, 162; as fortune, 65
xiang 象 (counterpart, image; sign), 180; in "Appended Statements," 41, 64, 107–8; counterparts as shadows, 94; of the female ruler, 205n14; as hexagrams or trigrams, 95, 129; in *ji xiong zhi xiang* 吉凶之象, 7; as sun or moon, 115; in *tianxiang* 天象, 5, 180, 213n8; in

xiang 象 *(continued)*
　titles of texts; 25, 196n24; Yao's tracing of, 68
Xiang Yu 項羽, 87–88, 214nn28–29
xiao 效 (verifying event, verify), 107, 129, 144, 147, 164, 181, 228n82; as appear, 154; as evidence, 133
Xiaojing 孝經. *See Classic of Filial Piety*
"Xibo kan Li" 西伯戡黎. *See* "Earl of the West Conquers Li"
"Xici" 繫辭. *See* "Appended Statements"
"Xingde" 刑德. *See* "Harshness and Favor"
Xiongnu, 50, 215n34
Xu Zhentao 徐振韜, 24, 195n15
xuanji yuheng 璿璣玉衡 (Agate Orb and Jade Crossbar), 68, 209n74
Xunzi 荀子, 4

yan 驗 (verify, verification), 80, 107, 124, 181–82; in titles of texts, 7, 186
Yan Shigu 顏師古, 214n26, 220n50
Yan Ying 晏嬰. *See* Yanzi 晏子
Yang Xiong 揚雄, 79, 141, 170, 174, 187n4, 200n74
Yanzi 晏子, 3–5, 36, 47, 105, 175, 188n6
Yanzi chunqiu 晏子春秋 (Annals of Master Yan), 3, 188n6
"Yao" 要. *See* "Essentials"
Yao 堯: as one of five sage-kings, 208n49; in memorials, 115, 145, 225n39. *See also* "Canon of Yao" and Shun 舜
"Yao dian" 堯典. *See* "Canon of Yao"
Yates, Robin, 40
Year Star 歲星. *See* Jupiter
Yellow God 黃帝, 28, 55, 185–86, 205n14, 208n49

Yellow River Chart and the Luo Writings, 82, 150, 186
yi 異 (anomaly), 99, 170, 180; in memorials, 118, 124, 125, 136, 141, 156, 158, 162, 220n61. *See also zaiyi*
Yi Yin 伊尹, 72
Yijing 易經. *See Changes*
yin and *yang*, 47, 51, 57, 59, 64, 75, 80, 108, 109–10; 113–21, 128–29, 138, 145–50, 160, 165, 177–78, 179, 181
Yin Min 尹敏, 112–13, 218n24
Yin 胤, 71, 210n86
Ying Zheng 嬴正. *See* Qin Shihuang
Yu 禹, 71, 145, 208n49, 218n33, 225n39
Yu 虞 and Rui 芮, people of, 69
Yu Shi 雨師. *See* Master of Rain
"Yue ling" 月令. *See* "Monthly Ordinances"

zai 災 (disaster), 41, 60, 98, 180–81; in memorials, 112, 117, 119, 120, 130, 147, 160, 169
zaiyi 災/灾異 (disasters and anomalies), 80, 180, 213n11; in memorials, 108, 114, 124, 125, 132, 135, 136, 143, 160, 163
Zhai Fangjin 翟方進, 133–35, 221n78, 222n87
zhan 占 (sign-reading), 179; comets in "Wuxing Treatise," pertaining to, 98, 99; Kongzi's views on, 44; in titles of texts, 7, 25–26, 186, 196n22, 196n24; in *xingzhan* 星占 or *zhanxing* 占星, 12; Yang Xiong, views on, 141; in *zhan hou* 占候, 200n81; Zhang Heng, views on, 228n82; in *zhanyan* 占驗, 80, 182
Zhang Fang 張放, 123, 226n47
Zhang He 章河, 160

Zhang Heng 張衡, 14, 230n3; apocrypha, views on, 162, 228n82; earthquake of 133 CE, 143, 161–64, 170–71; poetic works of, 176, 200n74, 231n6; seismograph, creation of, 162, 181
Zhang Huan 張奐, 130–31
Zhang Tang 張湯, 50
Zhang Xincheng 張心澂, 222n91
Zhao Feiyan 趙蜚燕, 225n46
Zhao Zhaoyi 趙昭儀, 225n46
"Zhaoli" 昭力, 40
zheng 徵 (proof; verify), 181–82, 228n82; in *jiu zheng* 咎徵 (proofs of blame), 125, 160, 162; in Many/Five Proofs 多/五徵, 111–12, 154, 226n51; as verify, 114; as *zhi* (musical note), 55

Zheng Xing 鄭興, 114–16
Zheng Xuan 鄭玄, 68, 120, 209n74, 210n82, 219n44
Zhou Gong 周公. *See* Duke of Zhou
Zhou Ju 周舉, 108–9, 120, 136, 138, 217n17
Zhou li 周禮. *See* Rites: Rites of Zhou
Zhu Chang 朱倀, 136, 138, 222n88
Zhu Fu 朱浮, 113–14
Zi Wei 子韋, 62, 132–37, 222n85
Zigong 子貢, 43–44
Zuji 祖己, 72, 123, 126, 211n93, 211n102
Zuozhuan 左傳 (Zuo Tradition), 4, 10–11, 87, 112, 114, 142, 199n61, 223n7
Zuyi 祖伊, 73–76